东南大学建筑学院国际联合教学丛书
International Joint Teaching Series of SEU-ARCH

POROs CITY
An Experimental Design Studio
for the 12th International Architectural
Exhibition / La Biennale di Venezia
2010.03–2010.08

孔洞城市
第12届威尼斯国际建筑双年展
受邀专题设计课程

张彤　Rainer Pirker　张慧　著

东南大学出版社
Southeast University Press

总 序

王建国
中国工程院院士
全国高等院校建筑学教育专业指导委员会主任
东南大学建筑学院前院长，教授，博士生导师

General Preface

Professor WANG Jianguo
Academician of Chinese Academy of Engineering
Chair of National Supervision Board of Architectural Education, China
Former Dean, School of Architecture, Southeast University

以践行"国际化"办学为宗旨的国际化教学正成为中国许多建筑院校新近最重要的设计教学趋势，这一趋势不仅包括了以往办学条件较好的诸如"老八校"和"新四军"之类的建筑院校，而且也覆盖到其他一些建筑院校。在我校举办的"2014年中国建筑院校境外交流学生作业展"中，参评学校已达35所。

从历史的角度看，国际化办学一直是世界名校建筑学办学的主要方式之一。美国哈佛大学、麻省理工学院和瑞士苏黎世联邦工业大学的建筑教育一直奉行的是全球延聘教授组织教学。中国早年的建筑教育也是一开始就与西方发达国家的建筑教学接轨。以中国建筑学办学历史最早的东南大学为例，当年的教材大纲、教程组织、教案编写等主要出自西方留学回来的教师之手，这些教师包括刘福泰、卢树森、鲍鼎、刘敦桢、杨廷宝、童寯、李汝骅、谭垣等，他们全都有着西方建筑学学习的经历。后来改革开放初期，也开展过一些外教教学的尝试，我本人就参加了由苏黎世联邦工业大学温克尔教授夫妇主持的建筑构成设计教学。而1946年创立的清华大学建筑系的奠基者梁思成也是美国学习背景。早期建筑设计教学的主要师资力量来自西方建筑教育的培养，所以在这个意义上讲，建筑设计教学的国际化也并不是全新的创举。只是在1950年代到1990年代，总体来说，改革开放前的中国建筑教育基本处于自我循环和封闭的状态。

Following the principle of "internationalizing" in architecture education, international joint teaching has recently become a trend among many architecture schools in China. Not only the schools ranked among "The Senior Eight" with more advanced education qualities, but also "The Junior Four", along with some other architecture schools, are chasing up with this trend. The number of schools participated in "Exhibition of Chinese Architecture Students Works of International Exchange, 2014", held in School of Architecture, Southeast University, already reached 35.

From a historic point of view, internationalized teaching is an ordinary phenomenon on among world-known architecture schools such as Harvard, MIT and ETHZ. These schools have always been employing professors worldwide for teaching. Actually, architecture education in China was in line with schools in western countries back in early years, taking the example of Southeast University, the architecture school with the longest history in China, the teaching program and curriculum of which at that time were all produced and organized by teachers graduating from western architecture schools, including LIU Futai, LU Shusen, BAO Ding, LIU Dunzhen, YANG Tingbao, TONG Jun, LI Ruhua and TAN Yuan, all of whom had overseas studying experience. Later in the early period of "Reform and Opening-up", we had also invited foreign teachers to our school as a teaching experiment. I, myself, had the experience participating in Architecture Composition Teaching Studio hosted by Prof. Heidi and Peter Wenger from Switzerland in 1983. LIANG Sicheng, who established School of Architecture, Tsinghua University also had studying experience in the U.S. Teachers involved in architecture teaching were mainly educated in the west in early years, therefore, internationalized teaching is not an innovation. Exceptionally, from 1950s to 1990s, China was basically self-circulating and enclosed.

今天，进入新千年的东南大学建筑教育走到了一个历史发展的转折点上，作为全国建筑教学的标杆，东南大学建筑教学必须应对当今全球建筑学领域学术的研究前沿和关注热点的流变。因此，我们将国际化作为新时期建筑教育努力突破的重点，而其中一个重要标志就是突破了以往多半教师先行出国学习进修，然后回校借鉴国际经验开展实验教学的做法，陆续开展了由境外教授和国际学生一起参与的、工作语言为英文的联合教学或工作坊项目。

多年来，东南大学建筑学院分别与美国麻省理工学院、加州大学伯克利分校、华盛顿大学、明尼苏达大学、德州农机学院和伍德布瑞大学以及瑞士苏黎世联邦工业大学、加拿大多伦多大学、荷兰代尔夫特理工学院、澳大利亚新南威尔士大学、奥地利维也纳理工大学和新加坡国立大学等合作组织了国际联合教学并取得显著成果。

经过多年的实践和持续积累，我们积累了较为成熟的国际合作办学和联合教学的经验。目前，东南大学建筑学院每年均开展6～8次国际联合课程教学，与国际知名建筑院系实现了校际学分互认，双授学位工作也在进行中。同时，东南大学建筑学院已经具备国际公认的办学特色和人才培养水准，拥有稳定和富有实效的国际联合培养的合作渠道，每年有一定数量的本科毕业生和研究生到国际知名建筑院系和规划设计机构继续深造和工作。

在教学实践中，我们也曾克服了不少实际的困难，如国内外学校的

Stepping into the new millennium, Southeast University is coming to a turning point. As a role model among all architecture schools in China, it needs to deal with the changes of academic frontiers and heat focuses of architecture education in a global scale. As a significant symbol, Southeast University steps beyond the tradition of sending teachers abroad for learning, who returns with foreign experience in order to conduct experimental teaching practice back home, rather, conducts series of international joint teaching programs or workshops involving foreign professors and international students, taking English as the official working language.

For years, our school has respectively carried out international joint teaching with MIT, UC Berkeley, University of Washington, University of Minnesota, Texas A&M University, Woodbury University, ETHZ, University of Toronto, TU Delft, University of New South Wales, TU Vienna and National University of Singapore, with significant outcomes.

After many years of practice and experience accumulation, we have gained mature experience in internationally collaborative education and joint teaching. So far, on average, our school carries out 6 to 8 joint teaching courses every year. We are also working on mutually recognizing credits and double degrees recognition together with world-known architecture schools. In the meantime, our school is internationally recognized by its education characteristics and qualities, having built up stable and substantial connections with world-known architecture schools for international joint education. Each year, a considerable number of undergraduate and graduate students continue their study in these schools or begin working in international institutions worldwide.

学期时段设置和教学计划安排存在的差异，不同文化背景的师生在合作交流时存在的价值观差异，以及教学经费筹措、教学活动管理、教学空间安排乃至师生的安全保险等等。为使这项教学活动实际可行，我们是从研究生阶段启动国际联合教学点，主要是教学时间和计划较为灵活，便于组织安排，同时研究生各方面较为成熟，境内外自主交流沟通和生活自理比较有保障。随着不断发展和经验积累，目前国际联合教学已经扩大到本科教学，课程设置和选题也有部分已经直接遴选在国外基地，让学生学习国外场地调研工作和人际沟通的能力。一直以来，东南大学建筑学院对国际联合教学工作中的教师人员配备给予优先，并在学院层面划拨了专门的经费加以支持。

国际联合教学极大地激发了学生的学习热情，使他们有机会直面国际化的教学授课环境，感受不同的教学传统、文化特点和创新活力，显著开拓了国际化视野，使学生们在日后深造和就业竞争中直接受益。而同时，国际联合教学对于教师也是一次参与和感受国际化教学环境的极好机会。

多年来，我们切身感受到国际联合教学对彰显东南大学建筑教育特色和优势的益处，此举也是东南大学建筑教育和办学国际化的重要组成部分。本丛书的陆续出版，从一个侧面见证了上述国际化教学方面取得的成果，由于很多工作仍然属于探索性的尝试，所以难免缺憾，希望读者批评指正。

We have also overcome many hurdles during the practice of joint teaching, such as the differentiations of academic schedules and curriculum between different schools, different perceptions of values of teachers and students from various cultural backgrounds, financing issues, management of teaching activities and teaching space, and even security insurance for each individual involved. To make these joint teaching programs more promising, at the beginning, we activated the teaching points only in graduate students' curriculum, since they have more flexible schedules for teaching organization, and they are generally more mature in every aspect, so that they can autonomously communicate with others and organize themselves. Along with the continuous development and experience accumulation, international teaching program has been extended to undergraduate curriculum. Some sites located abroad are chosen to offer better opportunities for students to learn the skills of site analysis and communication in foreign contexts. Our school has long been giving priorities to international joint teaching program, providing advanced teaching resources and extra financial support from school.

International joint teaching program greatly encourages students' enthusiasm of studying, offering them opportunities of facing directly to internationalized teaching environment, coming across with various teaching traditions, cultural characteristics, and creative energies, broadening their perspectives internationally, for their own benefits in future studies and careers. At the same time, International joint teaching program is also a perfect chance for teachers to participate into international academic community.

Over years, we have experienced the benefits brought by international joint teaching, which enhances Southeast University's characteristics and advantages in architecture education. International joint teaching forms an important part of architecture education in Southeast University. Publication of this series of books, from a side aspect, witnesses the outcomes of international joint teaching mentioned above. Since a lot of work is still under experimental practice, some regrets are hard to avoid, any correction and comment are sincerely welcome.

目录

FOREWORD

Page 01

前言

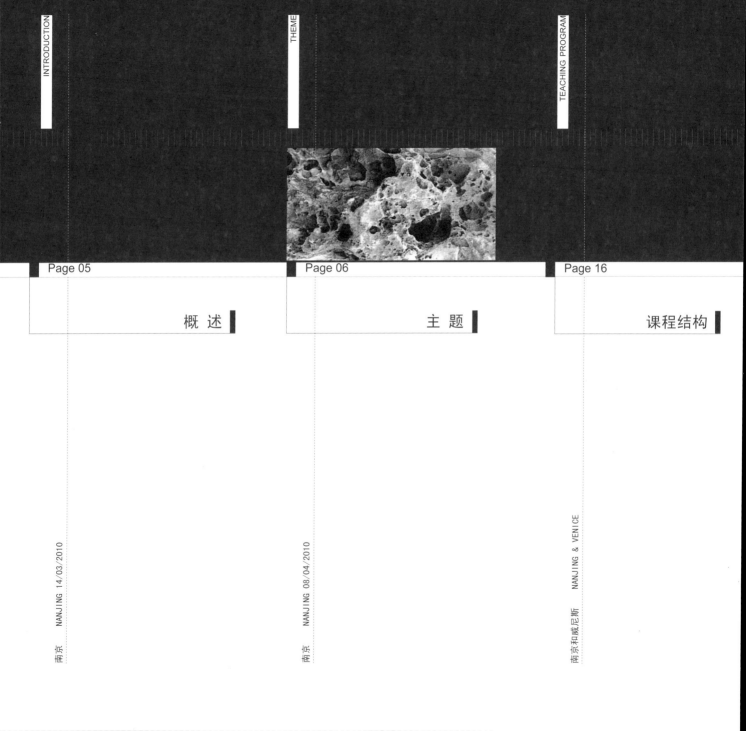

INTRODUCTION	THEME	TEACHING PROGRAM
Page 05	Page 06	Page 16
概 述	主 题	课程结构
南京 NANJING 14/03/2010	南京 NANJING 08/04/2010	南京和威尼斯 NANJING & VENICE

目录

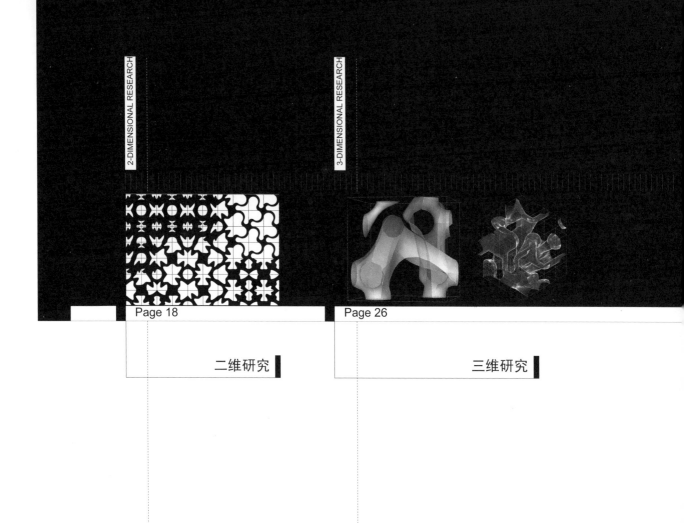

2-DIMENSIONAL RESEARCH

3-DIMENSIONAL RESEARCH

Page 18

Page 26

二维研究

三维研究

南京 NANJING 12/04/2010

南京 NANJING 15/04/2010

STUDY ON MATERIALS

FABRICATION OF THE EXHIBITION MODELS

Page 41

材料研究

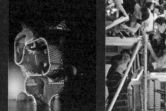

Page 42

展出模型制作

南京 NANJING 18/05/2010

南京 VENICE 05/07/2010

目录

ENCASEMENT & TRANSPORT

ASSEMBLY OF THE EXHIBITION MODELS

Page 57

Page 58

装箱与运输

实物模型组装

南京 NANJING 12/08/2010 & 13/08/2010

威尼斯 VENICE 15/08/2010

CITY WANDERING

EXHIBITION

POSTSCRIPT

Page 60

Page 67

Page 72

城市漫游

展览
第 12 届威尼斯国际建筑双年展

后记

南京与威尼斯 NANJING & VENICE 10/08/2010 & 17/08/2010

威尼斯 VENICE 27/08/2010—21/11/2010

前言
国际合作中的教与学
张 彤

Foreword
Teaching and Studying in International Collaboration
ZHANG Tong

与其说是前言,不如说是个人的一点感想。

东南大学建筑学院国际化联合教学的全面展开是在2004年。那一年,我刚从建研所调入学院工作,虽说不是新教师,但是每天走进课堂,心里还是怯怯的。不知从什么时候开始,大学里已经没有了教授和助教的差别,"青椒"前面没有老教师,少有榜样的示范和经验的传递,"菜鸟"直接上火线。回想十几年来的教学工作,个人的学习和提高大部分有赖于持续开展的国际联合教学,和与我长期合作的几位优秀教师。"国际合作中的教与学",从个人的角度,"教",可分享的经验不多;"学",在此过程中确实获益良多,所以我想先说说"学"。

2004年春季,维也纳"进步建筑联合会"组织两位中青年建筑师/教师瑞纳·皮尔克和迈克尔·舒尔特斯来东南大学做联合教学。我与皮尔克先生配合,从此开始了与他的合作与友谊。课题与教学组织都是瑞纳设计的,课程的主题是"都市耕作",内容针对当时南京规划中的第一条地铁线与沿线的地铁站。课题的设定基于这样的认识:地铁站不仅是交通站点,更是撒向城市,"播种"更新与再组织的"种子"。在两个星期高强度的集中教学中,瑞纳围绕空间主题,寻找和确立每一个方案的空间原型,在地形中挖掘空间潜质的教学方法给了我深深的触动,受此启发形成的教学方法,在设计课上一直沿用至今。

This is more of my personal thoughts than a foreword.

School of Architecture, Southeast University, roundly started international joint teaching in 2004. In the same year, I was transferred from Research Institute to the School. I was not a new teacher by that time, though, every day I still stepped into a class with timidity. It was not known since when differentiations between professors and assistants vanished in universities. "Green hands" are sent to front lines without any demonstrations from role models or experience passed on to them. Recalling my teaching career for the last decade, my personal improvement relied a great deal on the International Joint Teaching and a few outstanding teachers I have been long working with. From a personal point of view, I have less to share on "teaching" experience, but have more to say about "studying", as I indeed benefited a lot during this process, therefore I shall start to talk about "studying".

In the spring of 2004, two Austrian architects / teachers, Rainer PIRKER and Michael SCHULTES, organized by Architecture in Progress, Vienna, came to our school for teaching. I was teamed up with Rainer, and since then, our cooperation and friendship started. Rainer set up the topic and teaching framework, themed "Farm the city: Urban intervention" focusing on the 1st Metro Line of Nanjing and the stations along. Topic was chosen based on such an idea that underground stations are not only transport facilities, but more as "seeds" spread in the urban environment seeding regeneration and reorganization. During a two-week highly intensified studio, Rainer impressed me by the teaching methods of focusing on spatial themes, searching for and establishing prototypes for every single project, and exploring potential spatial qualities from the view of topography. Teaching methods inspired by the studio have been in use for me until now.

六年之后,瑞纳受邀携其作品参加2010年第12届威尼斯国际建筑双年展奥地利馆的展出,他不加犹豫地选择在东南大学再组织一次联合教学,这就是本书展示的内容。他完全依靠个人的理解,将形成复杂形态的参数化进程解析成为看得见、摸得着的物形操作过程,用于解释一种高密度城市形态的理想模型。这是在有限条件下一次很高水平的教学,其中的理念和方法一直作为我个人理解当代城市问题和参数化设计与制造技术的重要参照。

十几年来,我参与的另一系列化的联合教学是与洛杉矶的伍德布瑞大学建筑学院合作开展的。这不是一个声名显赫的学院,然而从2006年至2012年的七年间,尼克·罗伯兹教授与前后参与其中的七位美方教师,真诚地带领他们的学生每年五、六月间来到东南大学,与我们的师生共同开展教学活动。罗伯兹教授相信迅猛变化中的中国城乡环境是传统建筑学从未面对过的问题,也是在大学教育中最有价值的研究对象。七年中,教学的内容从秦淮河沿岸的旧城更新到下关江边的工业设施再利用,从高度密集的城市结构到徽州农村的景观系统。七年的教学在我们的师生面前展现出"景观都市主义"这一新的学科方向宽阔的理论与融合性的方法。最为可贵的是,认识不是来自生涩的读解和生搬硬套,而是在教学实践中鲜活的讨论与设计探索。这是与伍德布瑞大学连续七年联合教学最有价值的馈赠。

不仅如此,尼克·罗伯兹教授本人的真诚、尊重、耐心和儒雅给每一位参加联合教学的师生留下了深刻的印象,他展现了一位优秀教师的理想人格。在我遇到困难、感到烦躁时,有时会想起他。我相信,这种潜移默化的影响会存留在东南大学很多教师的心中。

以上是我作为一名一线教师参加联合教学的个人体会。粗略统计下来,十几年来在东南大学举办的国际联合教学超过了一百次,覆盖了建筑学院三个一级学科的全部专业方向,合作的学校包括了除非洲以外各大洲代表性的建筑院校。国际联合教学的方式由之前单纯请进来,到我们的师生走出去,在国际的语境中发现问题,寻找解决问题的方法。

Six years later Rainer was invited to exhibit his work at the Austrian Pavilion for the 12th International Architectural Exhibition in Venice. Without any hesitation he chose Southeast University to organize a joint teaching studio specifically for the exhibition, the outcomes of which form the contents of this book. Based on his personal understanding, he interpreted the parametric progress of complex generation and transformed it into tangible operational process, which can be seen and touched, to create an ideal model of urban morphology in high density. This is a studio conducted under limited conditions, producing high quality outcomes, the teaching concepts and methods of which have ever since become an important reference for me to understand contemporary urban issues, as well as parametric design and manufacturing.

During the last decade, another series of international joint teaching program I have participated in was collaboration with School of Architecture, Woodbury University, Los Angeles. This is not a worldly famous architecture school, yet, from 2006 to 2012, for seven years, Professor Nick ROBERTS and up to seven his colleagues brought their students to Southeast University every May or June to carry out this studio course together with our students. Nick believed that rapid changes of Chinese urban and rural environment had raised up many questions that traditional architecture education never faced before, which should breed most valuable research subjects in university. Within those seven years, studio topics ranged from "Urban Regeneration along Qinhuai Riverbanks" to "Re-utilizing Industrial Facilities in Xiaguan", from "Urban Structure of High Density" to "Landscape System of Rural Huizhou". Seven years of serial studios had demonstrated the broad and flexible theoretical framework and confluent methodology of emerging "Landscape Urbanism" for SEU students. The most precious part of this series studio was that recognition did not come from awkward reading of theories, rather, was gained from lively discussions, design and making. It was the most valuable gain received during the seven-year joint teaching program with Woodbury University.

Moreover, Professor Nick ROBERTS' sincere, respectful, patient, and gentle characteristics impressed everyone involved in the studios. He demonstrated an outstanding personality being a competent teacher. Sometimes when I encountered difficulties or felt lost, I would think of him. I believe this unconscious influence will remain in many teachers' minds at our school.

Above is my personal experience as a teacher taking part in international joint teaching courses. Within the last decade there have been roughly more than 100 international joint teaching studios carried out at Southeast University, covering all research directions of the three first-level disciplines. The schools we collaborated with include representative architecture schools from all continents except Africa. Collaboration methods change from "inviting in" to "going out", finding problems in international contexts and searching for solutions.

东南大学建筑学院内绝大部分年轻教师都参加过联合教学，参加的学生由前些年以研究生为主，到本科生、研究生并重，国际化的教学也由之前集中设置的工作坊发展到进入本科与研究生的日常教学体系。

2004年以来稳定而常态化的国际联合教学，伴随着东南大学建筑学院在国际化语境中的学科成长，成为开阔视野、培养人才、培育学科增长的重要途径。走过这些年，之所以要把"国际联合教学"单独拿出来回顾和总结，是因为十几年以前，"国"的界限是如此生硬，以至于遮蔽了彼此的视野，桎梏了思考和判断，造成了那么多的偏见与误解。教学确是最具实效性的交流方式，因为面对学生我们无可躲藏。国际联合教学以最鲜活的方式让界限内外直接碰撞，期间产生的诧异、困惑、感悟与理解让我们彼此受益。

在这个全球化的时代，超越和消解各种不必要的人为界限本身就是教育的宗旨。"国际化"不是最终的目的，也不应成为特别的方式和状态。在东南大学，"国际联合教学"作为特定时期的特定方式，其形式和作用也都发生着改变。驶出狭长内河的舟楫，在出海口，更需要辨析自己的方向与航标。如果说前十年的国际化办学主要为了打开边界、跨越壁垒，那么今后的目标则是在已然常态化的国际语境和开放平台上，更加鲜明地确立自己的学科特色和办学方向。

一百多年前，张之洞在《劝学篇》中提出"中学为体，西学为用""举国以为至言"。如今，中西二元对立已不再成为必然，学术发展日益面对全球问题，思想体系日趋交融。在这样的时代，教育的本体自觉显得尤为着重。树立自我的教育理念，明确人才培养目标，以我为本，主动选择和利用国内外教育资源，在全球化的思想语境和评价体系中，赋予这个源远流长的传统学科新的价值。唯有如此，我们才能通过回顾这十年的"青葱岁月"，看清"国际联合教学"的开拓性贡献。

Most young teachers in our school have experiences in these courses. In early years, students were mainly composed of graduate students, but now, of undergraduate and graduate students in equal numbers. Internationalized teaching developed from intensified workshops to being part of the regular curriculum of undergraduate and graduate teaching.

Since 2004 stable and regular practice of international joint teaching, along with the development of major disciplines, has become an important approach to broadening students' perspectives, educating students and improving the disciplines. The reason to bring up international joint teaching for reviewing and concluding is that, over ten years ago, the national boundaries used to be so harsh that they blinded each other's perspectives, limited people's thinking and judgements, resulted in so much prejudice and misunderstanding. Whereas, teaching is the most efficient communicating method, because we have nowhere to hide ourselves from students. International joint teaching, in a most stimulating way, lets the worlds both inside and outside boundaries collide, the astonishment, confusion, inspirations and understandings happening during the collision benefit everyone.

When we live in a globalized era, with the education targets of surpassing and eliminating any unnecessary manmade boundaries. Globalization itself is neither an ultimate purpose, nor should it be a special method or condition. At Southeast University, international joint teaching, as a specific method in specific period, its form and function also change. A ship sailing out of narrow inland waterways needs to discriminate its directions and navigation marks. If for the past decade, our aim of internationalized education was to open up boundaries and to overcome barriers, then, from now on, this aim will become to establish our own principles' characteristics and education direction more distinctively, on an open stage, within the already regularized international contexts.

Over 100 years ago, in his article titled "The Encouragement of Study", ZHANG Zhidong raised up the concept of "taking western techniques for practice on the mainstay of Chinese study", which was "worshiped as wisdom all over the country". Nowadays it is not necessary to polarize the Chinese and western cultures. Academic development faces more and more global issues. Theoretical systems are interweaving every day. Recognition of education autonomy seems to be more critical. We shall endow this old profession a new value by establishing self-awareness in education ideologies, clarifying cultivation targets, based on our own benefits, initiatively choosing and using both national and international teaching resources, within the globalized theoretical contexts and evaluation system. Only then, can we recognize the pioneering contribution of international joint teaching by reviewing its progression over the past decade.

概 述

受奥地利教育、艺术与文化部及其策展人艾瑞克·欧文·莫斯的邀请,奥地利建筑师、东南大学客座教授瑞纳·皮尔克与东南大学建筑学院联合开展设计课程"孔洞城市",成果参加第12届威尼斯国际建筑双年展奥地利馆的展览。课程教学开始于2010年3月,由建筑师瑞纳·皮尔克与东南大学建筑学院的张彤教授、张慧博士、虞刚博士联合指导,12名东南大学建筑学院的硕士研究生参加。教学的内容针对当前日益迅猛的城市化进程所显现出来的问题,以空间的密度及其结构为主题,以模数化设计和制造为方法,在抽象的模型层面上研究发展一种城市空间结构,区别于当今城市以分离个体的累加形成蔓延,这种新的模型通过空间的掏挖形成密度和整体的结构。教学成果最终表现为三件以"孔洞"为主题的大型模型。2010年8月,全体师生携教学成果赴意大利参加威尼斯国际建筑双年展。

组织机构:
奥地利教育、艺术与文化部
东南大学建筑学院
课程指导:瑞纳·皮尔克、张彤、张慧、虞刚
软件指导:卡特瑞·塔玛若
课程成员:
顾 鹏、胡 博、李沂原、徐臻彦
曹 婷、杨 晨、杨文杰、周艺南
厉鸿凯、李珊珊、史 晟、韦栋安
志愿者:
刘奕秋、吕一明、睢家俊、杨浩腾、姚远、张杰亮
部分照片由耿涛博士拍摄。
模型制作得到东南大学建筑学院造型实验室的协助。

INTRODUCTION

Invited by Eric Owen MOSS, curator of the Austrian Pavilion, and the Austrian Ministry of Education, Art and Culture, School of Architecture, Southeast University (SEU-Arch), together with Rainer Pirker (Austrian architect and Guest Professor at Southeast University), exhibited the accomplishments of design studio "POROsCITY" in the Austrian Pavilion as a contribution to the 12th International Architectural Exhibition/La Biennale di Venezia 2010. As a joint teaching program specifically designed for the event, the studio "POROsCITY" was started in March 2010. The program was supervised by Architect Rainer PIRKER and SEU-Arch Prof. ZHANG Tong, Dr. ZHANG Hui and Dr. YU Gang. Twelve graduate students participated in the program. To challenge the contemporary problems of radical urbanization, the teaching program chose spatial density and structure as the theme, and applied modular design/manufacture methods to develop a sort of urban spatial organization by abstract models, which differed from the spread of detached units accumulation in nowadays city. The new model forms density and integral structure by space excavation. The achievements of the studio were presented at Austrian Pavilion by three models with the topic of "POROsCITY". All the teachers and students went to Venice and attended the 12th International Architectural Exhibition in August 2010.

Organizers:
Austrian Ministry of Education, Art and Culture
School of Architecture, Southeast University
Supervisors: Rainer PIRKER, ZHANG Tong, ZHANG Hui, YU Gang
Software Supervisor: Kadri TAMRE
Studio Members:
GU Peng, HU Bo, LI Yiyuan, XU Zhenyan
CAO Ting, YANG Chen, YANG Wenjie, ZHOU Yi'nan
LI Hongkai, LI Shanshan, SHI Sheng, WEI Dongan
Volunteers:
LIU Yiqiu, LV Yiming, SUI Jiajun, YANG Haoteng, YAO Yuan, ZHANG Jieliang
Partly photographed by Dr. GENG Tao
Supported by Model Laboratory of School of Architecture, Southeast University

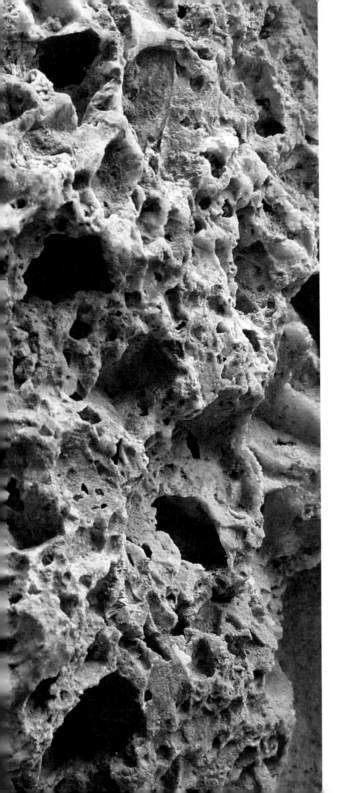

主 题

课程教学的主题"孔洞城市"中包含了两个核心的概念——"孔洞"与"城市"。

THEME

POROsCITY, the theme of this teaching program, embodies two core concepts – POROSITY and CITY.

孔 洞

孔洞是一种普遍存在于自然界和人工世界的结构。

孔洞描述的是空间，而不是物体。

孔洞表现的是具有整体性的结构，而不是离散喧闹的个体。

孔洞是在一个已经存在的整体中掏挖的空间，而不是分离个体的松散累加。

在方法层面上，孔洞是减法，而不是加法。

孔洞寻求与阳光、空气以及流动性相平衡的密度。

孔洞产生于模数化的单元及其逻辑的组织结构。

孔洞具有生命的机制、集约的密度、广泛的适应性和可持续性。

POROSITY

POROSITY is a universal structure existing in both the natural and the artificial world.

POROSITY is to describe the space, not the object.

POROSITY is to present an integrated structure, not scattered individuals.

POROSITY is to hollow out space within an existing whole, not to accumulate separated units.

POROSITY is a deductive approach, not an additive one.

POROSITY is looking for a certain density that may keep daylight, air and circulation in balance.

POROSITY is created with modular elements and their logical combination.

POROSITY is provided with organic mechanism, compact density, comprehensive adaptability and sustainability.

瓦德的绿洲镇，阿尔及利亚
Oasis Town, EL Oued, Algeria

城　市

城市汇集了地区的意愿、力量、限制与可能性。

CITY

A city is the locus of will, power, restrictions and possibilities of a region.

原始聚落

最初的聚落几乎无一例外地表现为以简单的个体通过高效的组织结构形成密集均质的整体。它们体现了人们用原始的技术手段适应自然环境，以最小的代价获取生存条件的智慧。虽然低下的经济技术水平决定了聚落的有限规模，但是这种单质的平向结构本质上是无限蔓延的。

PRIMITIVE SETTLEMENT

Almost without exception, every original settlement was set up as a compact, homogeneous unity, organizing simple individuals by structure with high efficiency. This phenomenon reflects human wisdom that people could adapt to natural environments utilizing primitive techniques, and meeting survival requirements with minimized cost. Although the low technical and economic levels at the time limited the scale of those settlements, fundamentally such a simple, homogeneous, planar/ horizontal structure could spread indefinitely.

巴黎 Paris

君政城市

文明的进步使社会产生了等级和秩序,均质聚落的平向结构被破坏,城市开始显现竖向性,具有了中心和边缘。城市的形态物化了社会内在的力量和组织秩序。帝国时代的罗马显现出君王强大的意志以及从贵族到奴隶的社会等级;在19世纪中叶的巴黎,短短20年时间里,拿破仑三世的个人意愿和强盛的君主国家体制创造了人类历史上最为恢宏的城市景观。

CITY OF MONARCHISM

With the progress of civilization creating rank and order in a society, the planar/horizontal structure of homogenous settlements was destroyed and cities began to develop vertically, with centers and edges. City configurations reflected social power and organizational order. For example, Imperial Rome incarnated the mighty will of sovereigns and represented social rank ranging from nobles to slaves; in mid-19th century Paris, over the short span of only twenty years, the personal will of Napoleon III and the monarchy system created one of the most magnificent urban landscapes in human history.

资本城市

进入工业社会以后,资本主义城市的形态清晰地对应于新型的社会功能和以资本地产金融为主导的经济机制。1920年代开始,高层建筑的建造成为资本力量的象征。在均质的网格地块上耸立起来的摩天楼迅速改变了纽约的天际线。

CITY OF CAPITALISM

In the industrial era, the morphology of cities in capitalist countries clearly reflected the new social programs and economic mechanism led by real estate finance. Since the1920s, high-rise buildings have been erected as symbols of financial might. Skyscrapers built on even grids rapidly changed the skylines of New York City.

左:巴塞罗那 Barcelona
右:纽约曼哈顿 Manhattan, New York

有机城市

与此同时,亚洲和拉丁美洲的很多城市,其发展仍然保持了自下而上的非规划的力量。城市的形态不具有宏大的景象和象征性的图形,其内部充满了丰富复杂的非正式结构,它们自发形成而且不断变化,具有生命肌体的特征和新陈代谢的机制。

CITY OF METABOLISM

Meanwhile, many Asian and Latin-American cities still develop from the bottom without planning. With no grand images or symbolic patterns, the city configurations have abundant intricate and informal structures, originating spontaneously and ever changing. These cities reveal an organism's character and the mechanism of metabolism.

东京新宿 Shinjuku, Tokyo

上：托斯卡纳卢卡 Lucca, Toscana
下：南京南捕厅街区 Nanbuting, Nanjing

物体城市／孔洞城市

除了历史学的类型认识，我们的课程倾向于从以下视点建立对城市的理解。

传统的欧洲城市产生于建筑物之间的关系，传统的中国城市产生于不同领域之间的关系。其共同的内质是将个体联系成为整体的致密的结构。密度在这些历史城市中具有首要的意义，它产生于紧密的社会结构、互为关联的社会生活、防御和节省土地的需求，也产生于对阳光和清洁空气的平衡考量。

与此相反，战后北美城市的发展产生的模式则是基于单个个体的累加和城市区域的无节制蔓延。这种被称为"物体城市"的模式导致了城市景观的碎裂。"曾经一度连续、和谐的城市景观现在被无关联的形体张扬设计和不断增高的摩天楼的罗列取而代之。城市肌理被割裂、被铲除。终极物体城市是无肌理城市。"（《物体城市》，张永和）

"……日照间距、容积率、建筑覆盖率、绿地率、退红线等规划指标对物体城市的形成起了关键作用。目前 1.6 以上的日照间距、常在 4.0 以上的容积率、同为 30% 左右的建筑覆盖率和绿地率、至少 5 米的退红线都在阻止城市获得密度，强迫建筑拉开距离，支持的只是物体建筑的建设，其结果也只能是物体城市而不是其他任何城市的出现。物体城市不是偶然的，是计划的，预谋的。物体城市是反城市肌理的城市理想的实现。

……

物体不会自己组成一个城市。公共交通系统不足或失败、交通堵塞、社区破碎或封闭、城市空间不明确、缺乏公共空间与步行街道、缺乏真正为低收入居民的公共住宅计划、缺乏商业购物网络等等，都是物体城市的典型症状。都市生活的快感失去了。最终导致城市机能的普遍瘫痪。"（《物体城市》，张永和）

这种以个体的累加为模式的城市化进程是松散的、低效率的。城市像一个畸形的肌体，不断地繁殖、蔓延，造成自然资源的极大浪费，对环境造成无法挽回的损害。

建设中的迪拜 Dubai under construction

CITY OF OBJECTS / POROsCITY

In addition to the study of historical typology, our teaching program intended to establish an understanding of the city from the following points of view.

Traditional European cities are based on the relationship of buildings while traditional Chinese cities are based on the relationship of territories. Their common feature is a compact structure integrating individuals into a whole, i.e. density is of principle significance to these historical cities. Density is generated from/by close social structure, interactive social lives, the needs for defense and land utilization, as well as the equal consideration of the needs for daylight and clean air.

On the contrary, the post-war model starting from North America is based on the accumulation of individual units and immoderate urban sprawl. This model, called "City of Objects", led to the disintegration of urban landscape. "The once uniform urban landscape is now a metropolitan field that is inundated with skyscrapers and individually expressive architectures with a disorganized or damaged fabric or simply without." (City of Objects aka City of Desire, Yung Ho CHANG)

"…Daylight Distance, Floor Area Ratio (FAR), Building Coverage, Greenery Coverage, and Setback are of paramount importance for forming city of objects. With current Daylight Distance at 1.6 and up, FAR rising to 4.0 or more frequently, both Building Coverage and Greenery Coverage around 30%, plus 5-meter minimum setback, all these prevent city from gaining density, and force buildings separation, only the building construction is supported. There can be no other possibilities but a City of Objects. Therefore, a City of Objects is envisioned as intention and deliberately planned,and it's the urban ideal of anti-urban texture.

……

Objects do not automatically make a city. The conceived and/or compromised public transportation systems, traffic congestion, discontinuous neighborhoods, poorly defined urban edges, lack of public spaces and pedestrian streets, absence of substantial public housing program, and insufficient commercial infrastructure are among the typical symptoms of the City of Objects. Urban pleasure diminishes. The city as a mechanism and/or organism may potentially go from dysfunctional to defunct." (City of Objects aka City of Desire, Yung Ho CHANG)

Such urbanization based on an additive model is of loose structure and low efficiency. Such a city keeps growing and spreading like an abnormal organism, and causes vast waste of natural resources. Its damage to the environment is irreversible.

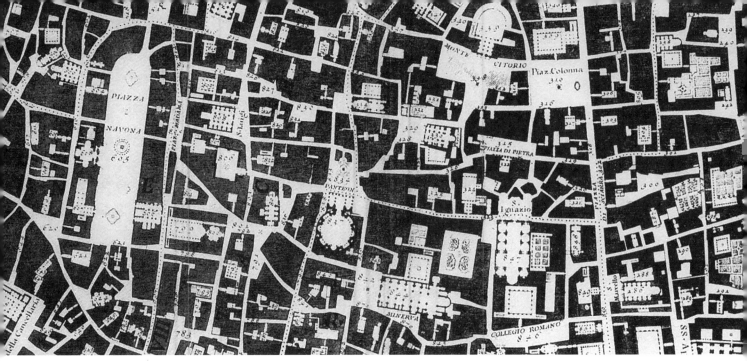

1748 年的罗马（诺利地图）　Roma in 1748 (map by Giambattista Nolli)

孔洞城市是与物体城市相反的概念。设想城市开始于一个密度为 100% 的整体体量，我们在其中创造孔洞，接受阳光和空气，建造基础设施，容纳公共活动。这是与个体叠加完全不同的过程，它是在整体中结构性的约减，在既有的整体中，通过孔洞获取密度。

与传统城市的二维网格不同，孔洞城市寻求一种三维的结构。在这个模型中，城市将在三维的空间体系中组织功能空间、基础设施、交通系统和绿化景观，在生态学意义上成为巨大的产生光合作用的系统。

孔洞城市不是针对某个城市的具体策略，而是在迅猛的城市化进程中，面临严峻挑战后的思考。到 2050 年，地球上的城市将容纳超过 75% 的人口。以个体的无序叠加为模式的消耗型增长将无法提供足够的扩展空间和生态支撑。我们必须寻求一种新的空间模式，在三维空间中创造结构和密度，以最为高效和节约的方式从环境中获取资源。孔洞城市作为一种抽象的实验模型，是新的城市理论的起点。

POROsCITY is a concept opposite to the City of Objects. Imaging a city as a mass with 100% density at the very beginning, in which we then create porosity to gain light and air, set up basic infrastructures and accommodate public activities. Quite different from the accumulation of individuals, this process is structural reduction, to achieve porosity within density by subtracting volumes from an existing entity.

POROsCITY is seeking a 3-dimensional model instead of the 2-dimensional grid of traditional Chinese cities. With this model, a city would organize functional space, basic infrastructure, circulation and greenery within a 3-dimensional spatial structure and become a large photosynthetic system in the ecological sense.

POROsCITY is not a specific strategy for a certain city, but the reflection on the serious challenges during the radical urbanization. In 2050, cities will have to accommodate more than 75% of the world's population. Cities as they are built today can not provide sufficient development space and ecological support for the consumptive growth. We must find a new spatial model to create 3-dimentional structure and density so as to obtain resources from the environment in a most efficient and restrained/frugal way. POROsCITY, as an abstract experimental model, is just the start of new urban theory.

模数系统与组合逻辑

在形式生成的方法上,课程要求以为数有限的单元体,通过逻辑清晰的组织方式,产生具有高度复杂性和丰富性的整体。因此,有效的模数系统和组织逻辑成为关键。中国的建构传统一直强调网格化和模数化,尽管在城市层面上大多表现为二维组织。在孔洞城市中,我们发展三维网格和模数系统。

由于模型最终要在威尼斯国际建筑双年展的现场组装起来,因此模数化又成为必需。它大大简化了制作、运输的过程。我们可以在南京完成所有单元体的制作,仅仅通过随身行李托运的方式,就可以化整为零地把单元体运到威尼斯。在那儿,我们只需要使用简单的工具,按照既定的组织逻辑和拼装方法,将它们组装起来。

MODULAR SYSTEM AND COMBINATION LOGIC

Concerning form generation, the teaching program intended to help students learn the methods to create a complete (whole, unified) object full of complexity and interconnectedness (in the sense of porosity), by logically diversifying the combinations of a limited number of elements. Therefore, effective modular systems and combination logic plays a crucial role. Despite two-dimensional spatial organization at the city planning level, traditional Chinese architecture always emphasize gridding (grids) and modularization. In POROsCITY, we are to develop three-dimensional gridding (grids) and modular system.

Moreover, since the models were expected to be assembled on site in Venice, modularization became necessary. It remarkably simplified and shortened the process of fabrication and transportation. In this way, all the elements could be fabricated in Nanjing and transported separately as check-in luggage to Venice. Then and there, only with simple implements, they were easily assembled according to the set logic.

课程结构 TEACHING PROGRAM

文献阅读 reading	理论教学 study on theories	城市 / 孔洞 / 预制和装配 city/porosity/fabrication and assembly
	孔洞研究 study on porosity	土地 / 密度 / 结构 / 自然结构 / 人工结构 properties/density/structure/ natural structure/artificial structure
软件培训 software	二维研究 2-dimensional research	平面单元 / 组合 / 图形 2D element / combination / pattern
	三维研究 3-dimensional research	单元体 / 组合 / 模型 3D element / combination / model
		单元体实物模型研究 physical models of 3D elements
	材料研究 study on materials	质感 / 重量 / 费用 / 运输 / 组装 / 防水 / 耐久 texture / weight / cost / transportation / assembly / water proof / durability
	展出模型制作 fabrication of the exhibition models	单元体和组装试验 elements fabrication and assembly test
	南京城市漫游 city wandering Nanjing	南京：东南大学校园、古城墙台城 Campus of Southeast University and the ancient city wall "Tai Cheng" in Nanjing
	装箱与运输 encasement and transportation	
	实物模型组装 assembly of the exhibition models	威尼斯公寓 apartments in Venice
	威尼斯城市漫游 city wandering Venice	威尼斯圣马可广场 Piazza San Marco in Venice

个人工作 individual work
小组工作 group work
数字模型 digital model
实物模型 physical model

南京工作阶段 working phase in Nanjing 2010/3—8　2010 年 3 月—8 月
威尼斯工作阶段 working phase in Venice 2010/8—9　2010 年 8 月—9 月

展览：第 12 届威尼斯国际建筑双年展
2010 年 8 月 29 日—11 月 21 日
Exhibition: the 12th International Architectural Exhibition
29/08/2010 — 21/11/2010

la Biennale di Venezia
12. Mostra Internazionale di Architettura

TEACHING PROGRAM

17

二维研究
2-DIMENSIONAL RESEARCH

课程教学的核心是以有限的不同的单元，通过逻辑性的组织结构，产生具有足够复杂性和丰富性的整体。为了更好地理解这种组织逻辑与孔洞的特性，学生们需要首先在二维体系中研究单元和单元的组合。练习从设计一组单元开始，反复将它们组合成为不同的图形，并且从可变性、丰富性、控制的可能性、密度、孔洞特性等各方面比较和评价这些图形。

通过对图形的检验，学生们需要反过来重新改进和发展单元的设计。这样的过程反复多次，不仅图形的质量和复杂性得到提高，学生们还能够认识到创造这样的图形所需的单元的特性，并且更好地理解组织的逻辑。这是一个试错的过程，不断地检验，不断地提高。

单元可以旋转和镜像。组织的网格可以基于方形、三角形、六边形或者其他几何图形，网格对旋转的可能性产生影响。尝试不同的网格，学生们可以认识到它们各自的优势和不足。但旋转过多，图形的变化反而减少，这也是需要比较和评测的。

The development of complex elements and structures for porosity should follow the logic of combination, in order to result in diversity and richness for the structure with a limited number of different elements. For a better understanding of this logic and the properties of porosity, the students were required to develop 2D patterns with a set of elements (1-6 elements) at first. For the improvement of these patterns, the students should start with one set (family) of elements, arrange them to patterns, re-arrange the same elements to different patterns and evaluate the created patterns concerning diversity, richness, potential of controlling, density, porosity property, etc.

After evaluation, a new family of elements was developed and the processes repeated as before. Doing this several times, the students can not only improve the quality and complexity of these patterns, but also understand the necessary property of elements for the creation of certain patterns, as well as the combination logic better. It is a trial and error method, with evaluation and improvement.

Elements can be rotated and flipped. The basic grid could be based on square elements, triangular elements, hexagonal elements or others. The grid has influence to the possibility of rotation. By testing different grids, the students learn about the advantages and disadvantages of different grids. One has more options for rotation, but maybe less options for variation of the patterns still. This has to be tested and evaluated.

杨晨的练习 Exercises by YANG Chen

设计的第一步是类似于平面拼图游戏的训练。工作方法是设计单元，将单元旋转拼接成为图案。基本的过程是设计单元、拼合组成图案、评价图案、修改单元，然后重复上述过程。评价是个困难的过程，面对抽象的图案，脱离了我们熟悉的具体属性，从日常的生活经验产生的审美不再适用了。目标是用最简单的单元拼合出最丰富的肌理变化。一段时间的尝试以后，我发现了二维游戏的规律：单元的复杂程度和图案生成的灵活性成反比。

The first step is training with exercises similar to puzzle game. Elements were designed, rotated and combined into patterns. Following the sequence - scheme elements, combine elements into patterns, evaluate patterns, improve the elements to a new generation and then go back to combination - the whole process was repeated continually. Amongst all the steps, evaluation is a difficult one. To those abstract patterns absent of concrete properties that we are familiar with, the aesthetic criteria derived from our daily life experiments don't work any more. The objective here is to create a most diversified patterns with simplest and least elements. After a long time of exercises and trials, I find the principle of this 2D game: the more complex the elements are, the less flexible it is to combine them into a pattern, i.e., complexity and flexibility vary inversely.

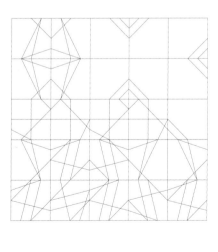

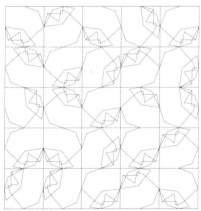

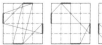

基于对单元连接点的观察分析，我得到了如下结论：控制点越少，单元通过组合形成的系统就能拥有更为多样的孔洞。如果每个单元内部的"半孔洞"的尺寸和形状都很类似和接近，单元组合后形成的孔洞也只能是乏味的均质。相反，只有加大单元内"半孔洞"的差异，才能增强最终系统的多样性和可能性。将4个基本单元按照90°、180°、270°和360°旋转就能获得16种衍变形态。我将最终的系统划分为3部分：第一部分，拥有较高密度并且孔洞的大小较为接近，这主要是通过选取了单元1、单元2、单元3这三种自身孔隙率较小的单元来实现的；第三部分，拥有低密度并且孔洞的差异较大，可产生较大的孔洞，这主要通过加入单元4这种自身孔隙率较大的单元来实现的；第二部分是一个过渡部分。由此，最终可得到一个具有渐变密度的孔洞系统。

Based on the observations and analyses on touching-points of elements, I got the conclusion: the less the touching-points are, the more the changes of the "void" inside the combination could be. If both size and shape of the "semi-voids" inside each element are similar, the "void" in the combination will be boringly even. The other way around, only if we enlarge the differences of the semi-voids of the elements, the porosity of their combination can gain more diversity and more possibilities. After 4 base elements were created, I rotated them by 90º, 180º, 270º and 360º, so I got 16 situations of the 4 elements. The final combination can be divided into 3 main parts: part-1 has higher density and less voids with similar sizes by means of combining elements with less semi-voids like element-1, element-2 and element-3; part-3 has lower density and more voids with higher diversity by means of adding elements with more semi-voids like element-4; part-2 is the intermediate state. Therefore, we got a result with gradually increasing porosities.

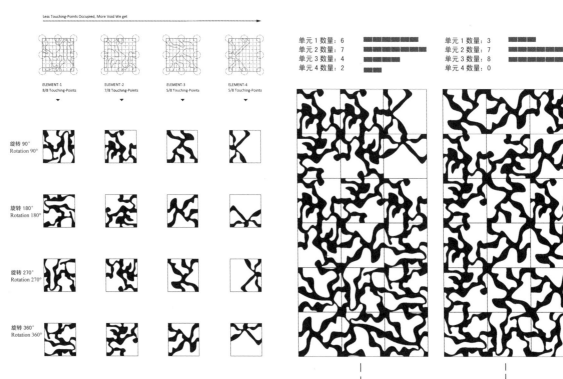

胡博的练习 Exercises by HU Bo

单元 1 数量：4
单元 2 数量：4
单元 3 数量：5
单元 4 数量：11

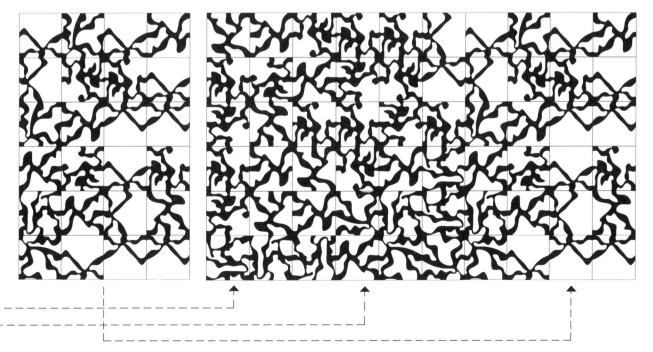

韦栋安的练习 Exercises by WEI Dong'an

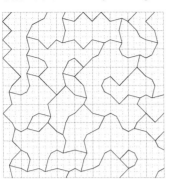

中点连接
connected at midpoint

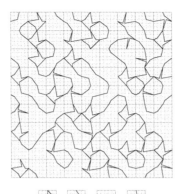

中点连接
connected at midpoint

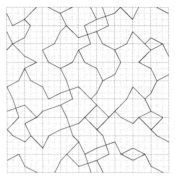

三分点连接
connected at trisection point

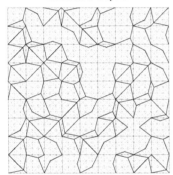

角部连接
connected at corner

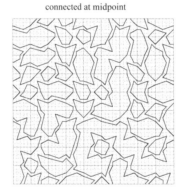

中点连接
connected at midpoint

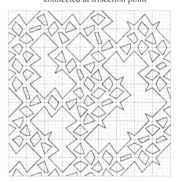

角部连接
connected at corner

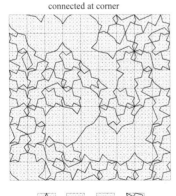

角部和中点连接
connected at corner & midpoint

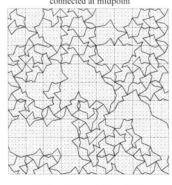

角部和中点连接
connected at corner & midpoint

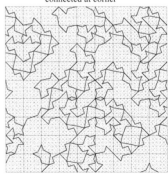

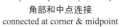
角部和中点连接
connected at corner & midpoint

二维研究是通过几个基本单元的组合形成异质的二维孔洞图案，具体操作有限制性要求：每个单元内必须是有图案的，既不能出现空的单元，也不能有实心单元；孔洞需依靠单元的组合形成。因此，孔洞本身的最大直径不会超过2个单元的基本尺度。要形成异质的孔洞密度，需要大小各异的孔洞。对单元进行更多次的划分可以获得小孔洞。由于连接点是确定的，因此自中心向外围的划分方式对于孔洞的异质性效果不明显，通过折线连接各个连接点的方法效果较好。

通过尝试不同的连接点和单元划分方式，2D图案的组织逻辑和孔洞规律更加清晰。简言之，要实现二维图案的孔洞异质性首先要对单元本身进行异质区分。

The 2-dimensional study is to contrive a 2D pattern with heterogeneous porosity by combination of several basic elements. It is required that neither blank nor solid element can be used, i.e. each element is some kind of figure, and the void area must be created by the combination of elements. So, the maximum diameter of the void is less than 2 basic elements'size In order to get a heterogeneous porosity, differently sized voids are needed. Smaller voids can be achieved by enhancing complication of the figure, i.e. more partitions within the element. Since connecting points (touching points) are fixed, the partition from center to periphery is not so helpful to the heterogeneity, and the way of using polyline to connect touching points has effect.

By changing the position and number of touching points and testing different partition way, we got to understand the combination logic of 2D pattern and find the porosity facts. In short, the heterogeneous porosity of 2D patterns is based on the uneven partitioning of the elements and different partition ways among the elements.

杨文杰的练习 Exercises by YANG Wenjie

密度与方向性研究
Research on Density and Direction

早期的二维练习揭示出方格网体系的若干不足：首先，方格网体系自身的复制会造成整体的均质性，要想获得变化丰富的图形，必须提高单元内部形态的复杂程度，这样又削弱了组合的多样性和灵活性；其次，正方形格网的图案发展方向为正交双向发展，总体形态的密度变化受制于格网体系；再者，正方形格网体系无法反映出一个综合复杂体（例如城市）的组织规则。

针对上述三个问题，我首先尝试了六边形网格体系，增加格网的发展方向，即从四向延伸发展为六向延伸；为了满足复杂体系的需求，我又进一步发展出三套格网叠加的组合式网格体系。相较于方形网格体系，多边形组合式格网体系更有利于创造复杂多变的图形。

Early 2-dimensional exercises exposed several disadvantages of square grids. Firstly, the repetition of square grids resulted in the homogeneity of the pattern. In order to achieve diversity within patterns, the complexity of the elements must be enhanced, which would weaken the diversity and flexibility of combination of the elements. Secondly, the patterns based on square grids can only grow in orthogonal directions, so that their density change is restrained by gridding system. Thirdly, square grids alone are not enough to reflect the organization principles of a complexity like a city.

In view of the mentioned three problems, I tried hexagon grids at first to obtain more growing directions for the pattern, i.e. from four to six directions. Then, to gain more diversity, I developed a combined system with three kinds of grids. Compared with square grids, such combined, polygonal gridding system could be more favorable to create complex patterns with the diversity of density/porosity.

网格组合研究
Grid Composition Research

图案拓扑
Pattern Topology

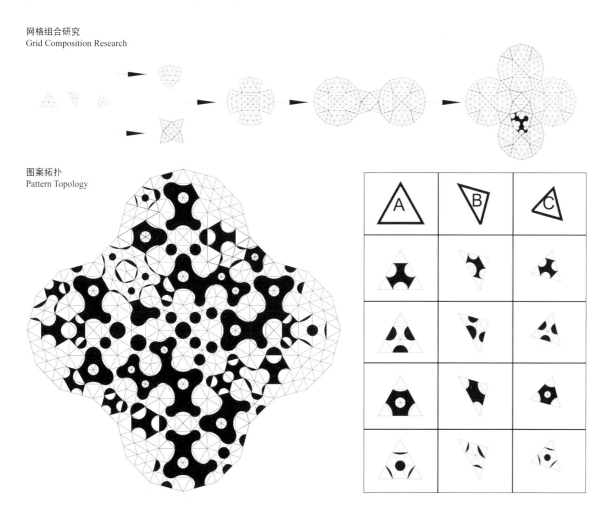

三维研究

从二维到三维的过程是一个新的挑战，涉及对于单元特性和组合方式的更为复杂的要求。在三维的空间中，孔洞具有了新的性状。任何在二维图形中出现的孔洞在第三维度上产生了复杂的空间中的联系。在对角线方向上的关系尤为重要。

3-DIMENSIONAL RESEARCH

The step from 2D to 3D is a new challenge with more complex requirements for the properties of the elements and their combination. Within a 3D space, we get new conditions of porosity. This means that the porosity represented in any 2D pattern could have complex connections of voids in the 3rd dimension. The relation on the diagonal direction is getting important.

为加强对三维关系的理解，学生须综合运用下列方法：

1. 试验性的实物单元模型

这是一个用小木棍或任何其他坚硬材料做成的简单的立方体框架，用于研究框架内空间的三维连接。立方体的边长在20~30cm之间。

模型是对立方体三维条件的简化模拟。学生们首先将二维单元的轮廓附在立方体的各个面上，这样他们就可以理解相互连接的关系（连接面）。然后他们可以用棍子、线、纸或纸板等材料来研究立方体中空间的几何形状，做出简单的三维工作模型单元。这些实物模型的研究是与电脑中的数字模型建构同时进行的。除了立方体各个面之间的关系，还要研究单元间的组合关系。

2. 剖面草图

关于剖面的草图能够帮助理解单元中的三维关系，确定单元的几何形状。从剖面草图中获得的空间意识将会十分有助于三维模型的建构。

3. 数字模型

在试验性实物模型和剖面思考的同时，学生们要在计算机上研究三维单元的数字模型，并且以一组单元反复研究组合关系。组合模型需要足够大，以便评价与孔洞城市相关的孔洞的特性。

4. 实物组合模型

在各种数字组合模型的比较之后，学生们选择其中一个制作实物模型。模型的制作需要达到基本的精确度，以便评测空间的质量、可变性与孔洞性。在这之后，学生们需要重新检验单元体和连接方式，做出修改，使之获得改进。这个过程需要反复多次，至少5次，最终获得有趣和有用的结果。重要的是，所有的修正应基于对检验和评价前次单元的合理标准。

To support the understanding of 3-dimensional relations, the students shall combine different methods:

1. Experimental physical model

This is a simple cube represented as a frame model made with sticks (wood or any other material being strong enough to use this frame model several times for the research of 3D relations within the cube). The size of this cube shall be 20~30cm for each edge.

This model shall be used to simulate with simple methods 3-dimensional conditions within the cube. The students shall use element contours of their 2D investigations on each face of the cube. By doing this they will first understand the relations of combination (touching surfaces). Then they use sticks, strings, paper and card board to think of geometry conditions of the cube and to develop simple working models of elements. These models are tools in addition to 3D computer modelling.

The conditions of combination can be studied as well as the relation of faces within the cube.

2. Section sketches

Section sketches could support the understanding of 3D relations in the cube and help to think of possible geometries of the elements. The awareness of 3D relations from section sketches will help the students to develop 3D models.

3. Digital model

Based on experiments in the experimental physical model and thinking in sections, students will develop 3D digital models of elements on the computer. A family (family A) of 3D element models will be used to set up combination models as well. Students shall develop several versions of combination (combination A1, A2, A3…)These combination models shall be large enough to evaluate the conditions of porosity related to POROsCITY.

4. Real physical model

Based on the evaluation of the digital combination models, the students shall choose one model to be built as a physical model. The models shall be accurate enough to evaluate the quality of space, variation and porosity. After evaluation of the physical model, the students start the process of improvement of the element (new family > B, C, D…) and the combination. To do this improvement, the students make the process from 1~4 as many times as they need to get an interesting and useful result. The process shall be repeated at least 5 times. It is highly important, that modifications are based on rational criteria from the evaluation of the element families before.

杨晨的练习：三维单元的研究　Exercises by YANG Chen: Study on 3D Elements

图形改进　Improve the patterns

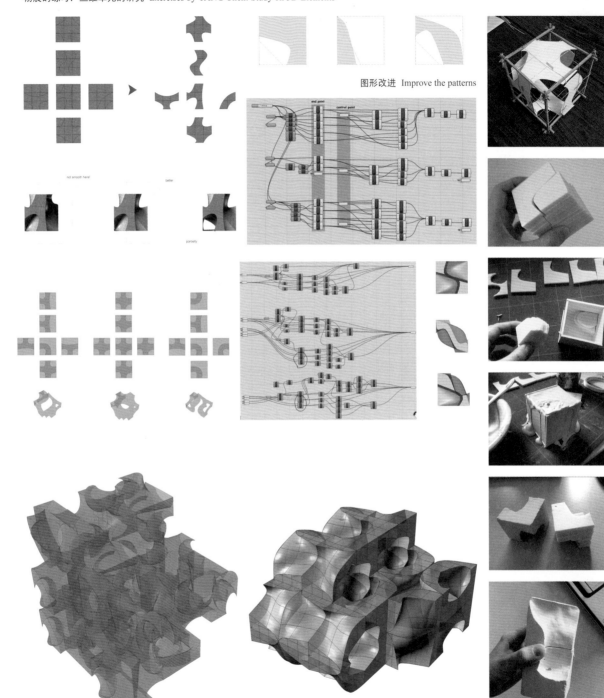

杨文杰的练习：三维单元及其组合密度研究 Exercises by YANG Wenjie: Study on 3D Elements' Combination and Density

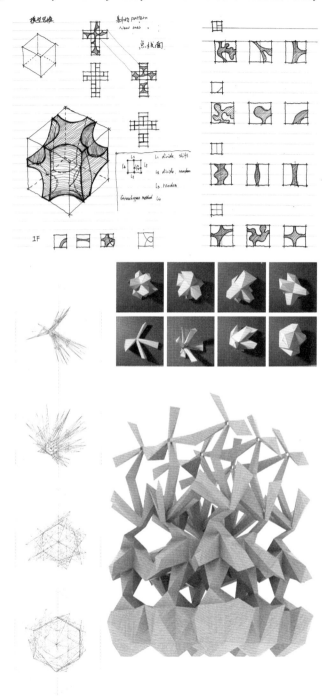

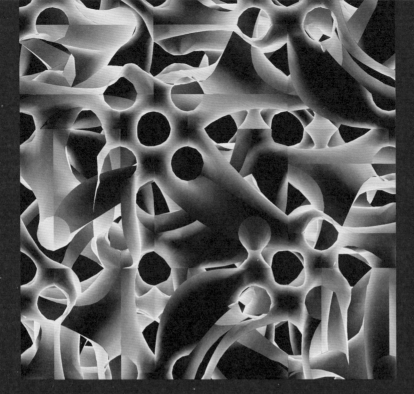

编组工作
数字模型的生成

在这个环节之前，学生们都是以个人为单位进行学习和研究的。在完成了对三维单元及其组合方式的研究之后，学生们被要求编组进行最终单元和组合模型的设计。学生个人获得了不同经验，改进了小组的工作。12名学生被分成了3组（每组4人）。小组工作的基本方式并没有改变，他们选择小组成员之前工作中最具发展潜力的模型，集中力量加强对其改进。在这个环节的研究中，学生们还要思考在单元中容纳不同尺度的孔洞，由此单元的复杂性将大大增加，而这必须基于学生们对复杂性已经具备一定的掌握能力。

通过小组的强化工作，最终的三维单元及其组合方式被确定。各组根据展现现场的要求，设计最大尺寸为80cm（宽）×93cm（深）×120cm（高）的最终模型，并要求在电脑中完成数字生成。

GROUP WORK
GENERATION OF THE DIGITAL MODEL

After the 3D element development of individual students, the students formed groups to work together for final design of elements and combination models. The different experience of the individual student could help to improve the work in the group. The size of each group was the same (4 students in each group). The basic process was the same as above. The work in the group intensified the way of improvement process. The complexity can be enlarged by integrating different scales into the development of elements. The step of scaling is only made, if the students can handle the complexity.
In the group work, final 3D models of elements and combination were developed. According to the physical conditions of the exhibition, each group developed the final digital result of a large scale object in the computer with the size no more than 80cm (L)x93cm (W)x120cm (H).

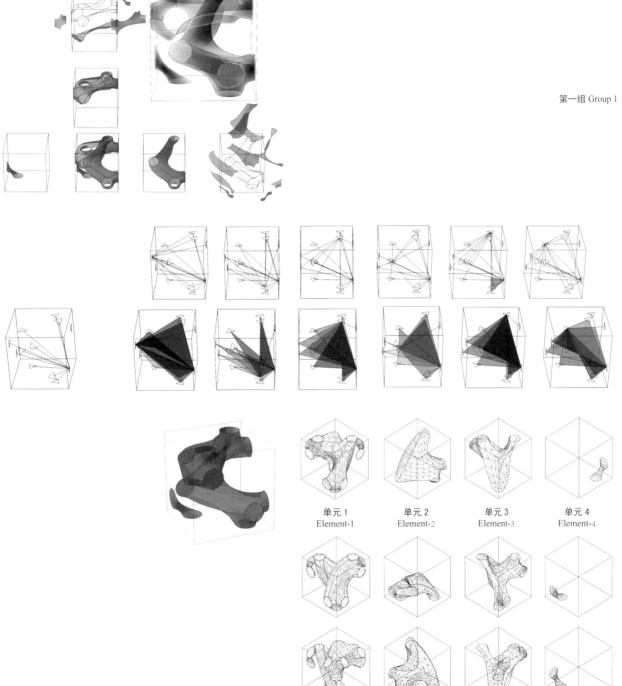

第一组 Group 1

单元1 Element-1　　单元2 Element-2　　单元3 Element-3　　单元4 Element-4

单元研究 Study on Elements

第一组 Group 1

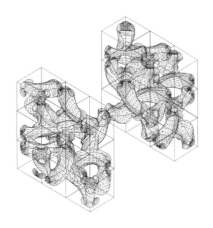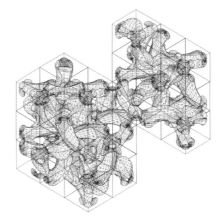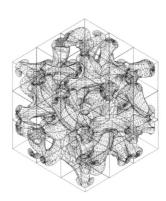

单元的组合 Combination of Elements

在三维建构的数字模型中，每个单元都极富变化，因此为整体提供了多样变化的可能性以及可识别性。单元的内部具有复杂交织的形态，能够造就一般的简单单元无法缔造的丰富整体。单元本身的复杂性使得组合的方式简单而清晰，组合后的形式丰富多样而超乎想象。最少种类的单元缔造出最复杂多样的整体空间是这个抽象的模型给予未来城市的启示。

Since each element is varied in our digital model, the whole model is characterized with identifiableness and various possible appearances. These elements with labyrinthine inner structure can build up a complicated entity that simple elements never achieve. The complexity of elements allows a simple and clear way of combination and the result of combination is variety and far beyond imagination. To achieve a most complex and diversified space with least types of elements, that is what this abstract model would tell the city.

第一组 Group 1

3-DIMENSIONAL RESEARCH

33

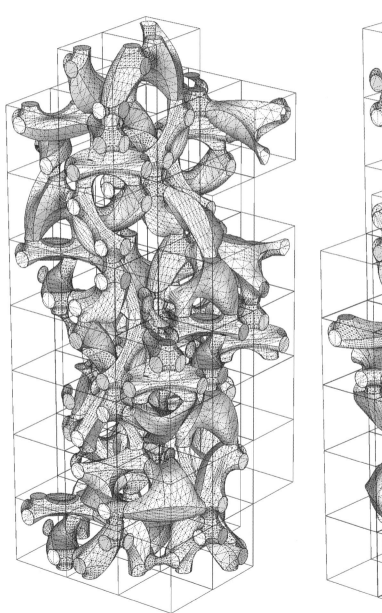
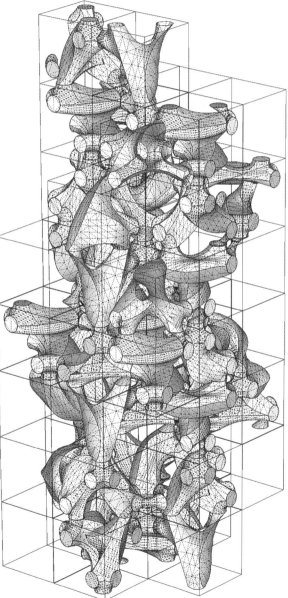

完成的数字模型 Completed Digital Model

第二组 Group 2

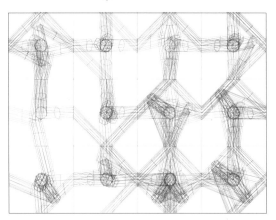

三维密度研究

密度概念的提出可以为形态和大小控制提供准绳，从而也为形态的调整提供了合理的依据。密度反映实体部分所占的比率，非实体部分即为孔洞。密度和孔洞同时体现在单元体和组合体两个层面上。如果将由若干单元体组合形成的整体设想为城市，密度能够贴切地描述自生长城市的构成与发展，因为各种内外因对城市的影响反映在城市各部分的密度变化中，这些影响因素促成了复杂的城市形态。

基于早期采用"枝杈系统"的初衷，我希望能够通过调节"枝杈"的密度来控制孔洞城市的变化节奏。受热带雨林上密下疏形态的启发，我将之反转，设计出密度逐层变化、上疏下密的三维体量。采用这种模式建构的孔洞城市能够使环境因素（日照、气流、雨水）由疏及密地浸润到城市底部，成为一种功能简捷、资源共享的理想城市。

STUDY ON 3-DIMENSIONAL DENSITY

Density is not only an important reference criterion while controlling the configuration and scale of entities, but also a reliable basis for the adjustment of the configuration. Density is the volume ratio of solid parts within an entity, hereby, the opposite notion is porosity. Density and porosity can be applied both at the element layer and at the layer of combined entity. Imaging a structure composed of several elements as a city, the constitution and the development of a self-growing city can be described relevantly with density, since the changes of density within a city reflect the influences of various internal and external factors that lead to complex morphology of cities.

With the original intention of "branching system", I attempt to adjust the density of "branches" to control the changes of my POROsCITY. Inspired by the tropic forest with dense branches on the top and loose branches or stubs near the ground, I reverse the branching system from "up-dense low-loose" to "up-loose down-dense" in the sense of density. A city with such "up-loose down-dense" branching structure could allow the natural factors such as sunshine, air flow and rain water to penetrate and infiltrate to the bottom of it; it would be an ideal city with clear functioning and resource sharing.

单元的组合 Combination of Elements

第二组 Group 2

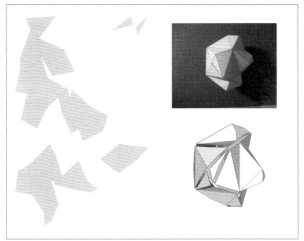

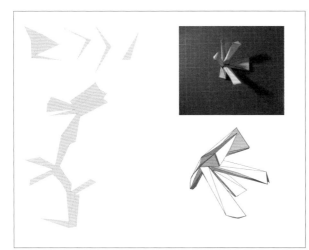
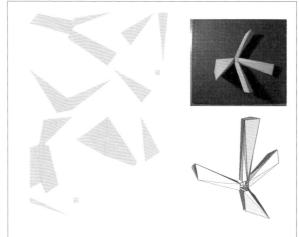
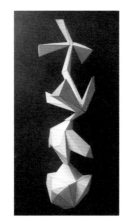
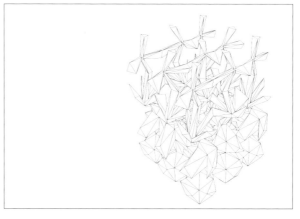

单元组合与密度研究
Combination of Elements
and Study on Density

第二组 Group 2

三维研究

我们抽象了丛林的枝杈结构，并使其物化为6种单元。每种单元都在三维空间内生成枝杈的结构，同时通过接触面与相邻单元相接，使得整体可以通过单元的对接不断发展，也诠释了城市的物质、能量、基础设施等相关系统连续流动的思想。经过不断地调整单元的位置关系、接触面的形态与个数、单元的旋转角度、相邻单元的类型，我们最终得到了理想的整体形态与空间效果。在设计中实体的密度自下而上逐渐降低来模拟丛林形态，从而在顶部为阳光、雨水等自然要素渗入城市提供充足的孔洞。在下层使用实体部分较多的单元，创造一个养分的集合，向冠部输送能量和物质，同时密实的实体能在结构上对上部形体提供支撑。

We abstracted forest branching structure and created 6 basic elements. Each element can develop to a branch in the 3-dimensional space and be connected to adjacent elements by touching faces. In this way, the whole entity can grow through the connection of elements, which also illustrates the idea of fluid on aspects of substance, energy, infrastructures and related systems in cities. After continual trials on the positional relationship of elements, the configuration and number of touching face, rotation degree of elements and the element types of adjacent units, we got the final model with the satisfying overall form and spatial effect. The bottom-up degradation of density simulated the forest structure could allow natural components like sunshine and rain water to penetrate into the whole city and provide the city with plenty porosity. The massive lower part collecting nutrient could deliver energy and substance to the upper part, and meanwhile, support the upper structure.

完成的数字模型 Completed Digital Model

第三组 Group 3

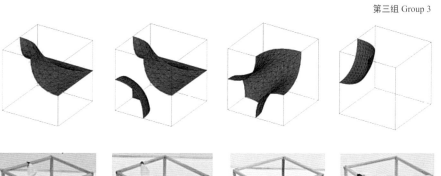

 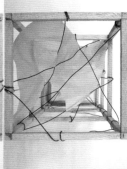 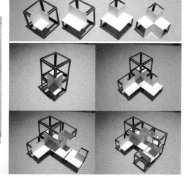

第三组 Group 3

三维研究

最终模型类似错综复杂的洞穴，但是真实尺度将比洞穴大数千倍。构建的方式是木片之间相互插接构成基本单元，基本单元的连接形成最终模型。组合逻辑采用的是完全对称的接触面，连续不断地制造出各种尺度的洞穴。密度的控制来自对单元数量的改变。我们能够写出新的脚本，使得这些单元自动拼合。这个脚本可以控制每个单元的出现概率。

孔洞城市是一种猜想，目的是给予人类更广阔的生存空间、更多样的生存方式。当然这样的城市背后一定有不同的社会结构和更加伟大的工程技术。什么样的社会需要这样的空间模式，什么样的人能接受这样的生活方式是一个值得探讨的话题。现在的城市都是来自于社会结构、政治制度、经济方式、自然环境。而孔洞城市不是，它是先于一切存在的。我们的模型完全不是为了再现一种幻想或者说启发一种发展方向，更不是解决一种看似不可能实现的工程技术问题。我们对解答某种答案没有兴趣，我们所做的是提出一个问题，激发社会变革和新的生活方式。

The real scale of our final model similar to a labyrinthine cave should be about thousands times larger than natural caves. Plywood slices inserted into one another form the basic elements that are connected into the final exhibiting model. Combination logic is based on the totally symmetric touching faces. Changing the individual number of each element can control the density of the whole. A new script controlling the emerging probability of each element was written to make the elements combined automatically.

POROsCITY, as a kind of suppose, is aimed at a more capacious living space and more various life styles. Certainly, such a city must be supported by a certain social structure as well as by great technology. Then, what kind of society would require such space structure? What kind of people can accept such a lifestyle? Different from contemporary cities based on social structure, political and economic systems, and natural environment, POROsCITY exists before all things. Our model is in no sense aimed to represent a fantasy or to suggest a direction, not even to solve technique problems. We have no interest in answering, but in questioning, in order to evoke social changes and inspire a new lifestyle.

第三组 Group 3

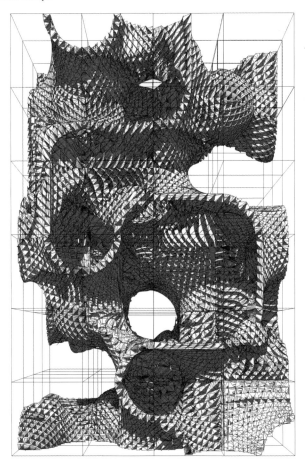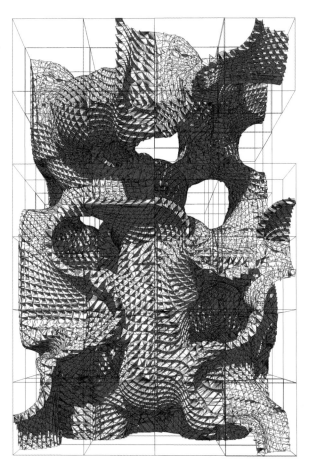

完成的数字模型 Completed Digital Model

材料研究

在基本完成了数字模型之后,实物模型研究的第一步是选择材料。各组根据自己的模型设计,选择可能的材料,并进行制作试验。他们要考虑的不仅是质感、色彩,还要考虑加工组装的工艺,制作的时间,选择尽量轻的材料以便运输,在室外展览的防水、耐久和坚固等要求。

STUDY ON MATERIALS

After the final digital models were mostly determined, the study on materials got started. The first step was to select a suitable material. According to its own design, each group chose one possible material for the physical model and tested its feasibility. Factors taken into consideration included not only texture and color, but also process and assembly requirements, fabrication duration, light weight for transportation, durability, firmness for exterior exhibition, and need to be waterproof.

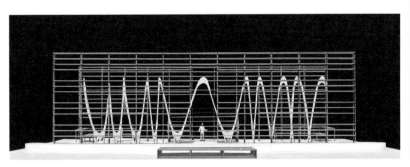

展出模型制作

FABRICATION OF THE EXHIBITION MODELS

策展人埃瑞克•欧文•莫斯将本届威尼斯国际建筑双年展奥地利馆的展览主题定为"建造中",意在表达奥地利成为当代世界建筑与城市规划思想汇聚的热点。展览的内容与形式颇具意象性,呈现了奥地利和国际建筑师、教师一起工作和教学的两面。

奥地利建筑师与教师在国外的工作被展示在展馆的外部,世界其他国家的建筑师与教师在奥地利国内的工作被展示在展馆建筑的内部,展览展示了建筑师和教师的工作。

瑞纳•皮尔克与东南大学师生的作品作为奥地利教师在国外学校的教学成果与其他13所学校的展品一起陈列于奥地利馆的内院中。东南大学是唯一入选展出的亚洲学校。

给东南大学提供的展台为240cm(宽)×93cm(深),要求展品高度不超过120cm,展台上放置了3个大尺度的"城市孔洞"模型。

"Under Construction" is the exhibition theme set by Eric Oven MOSS, the curator of the Austrian Pavilion at the 2010 Venice Biennale, to present Austria as an important centre for contemporary architecture and urban planning discourse. The showroom was designed symbolically to represent the two directions of how Austrian and international architects and educators are working and teaching together.

The exterior (the outer courtyard) presented the works of Austrian architects and educators working and teaching around the world, whilst the interior (the inner courtyard) presented the works of international architects and educators working and teaching in Austria. In all, there are works from architects and educators exhibited there. Showing the achievements of Austrian architects teaching abroad, the works of Rainer PIRKER and Southeast University (SEU) were placed in the courtyard of Austrian Pavilion. Displayed side by side with the productions from 13 other universities, SEU had the honour of being the only Asian university chosen.

SEU's exhibition area was 240cm in length, 93cm in width, with the height of works not to exceed 120cm. Three large models representing "POROsCITY" had set within this space.

南京：单元体与组合试件

三种富有质感的轻质材料和相应的制作工艺被选定用于展品的制作，分别是金属网/折叠绑扎、软木片/层叠、硬木片/织构。组装试验的过程中，单元体的连接构造方式得到改进。在软木片/层叠模型中，学生们在相接的软木单元体之间嵌入了木质插件，强化连接，增强整体的刚度；在硬木片/织构模型中，学生们反复试验交织木片的方向和厚度，在最大程度上获得面的连续感。组装试件的完成效果证明，三种材料的选择不仅体现了设计意图，材料本身以及相应的构造连接使得设计模型具有突出的表现力。它们或如薄纱般虚幻缥缈，或如地质层构般富有坚实的时间感，或在连续翻转的表皮间营造层出不穷的孔洞性。试件的成果令人兴奋和激动。

NANJING: ELEMENTS AND ASSEMBLY TEST

Three lightweight textural materials and matching fabrication methods were chosen for the exhibition models: wire mesh by folding and sewing, cork timber by layering, and plywood by gridding. Through a series of assembly tests, the tectonic methods to connect elements were improved. For the cork timber/layering model, plywood pieces were inserted into between the adjacent cork elements to strengthen the connection; for the plywood/gridding and weaving model, different directions and thickness of woven plywood pieces were tried in order to achieve ultimate continuum of surfaces. The success of the tests proved not only that the right selection of material successfully exhibited the design intent, but also that the material together with its matching tectonic combination endowed the models with remarkable expression. Some test results brought to mind chiffon, misty like a thin veil. Some awoke strong time-consciousness as strata layering. Others displayed unexpected, ever changing porous surfaces resulting from continuous turning. All results were very exciting and inspiring.

展出模型制作

第一组 Group 1

关于模型的特征如形式、颜色、肌理以及系统的孔洞密度和性质，我们试图给予最终展出模型与数字模型类似的属性——生物体般的有机形态，暗示城市如同生命肌体的生长法则。孔洞、复杂性以及多样性是设计的目标。幸运的是，暗黄色软木为概念的实现提供了理想的材质效果，这种材质能够让人联想起"孔洞"的特性。同时，足够轻的质量为飞机运输携带提供了可能。

We tried to make the eventually display model resemble digital model with properties like form, color, texture, system's porosity density and an essential character - organic form, in order to express our idea that a city could grow like an organism. Porosity, complexity and diversity are aimed at. Fortunately, sandy cork with the ideal effect for our idea was found. The appearance of this material seems associated with "porosity". Thanks to cork's light weight, it is easy for us to bring them to the exhibition by ourselves.

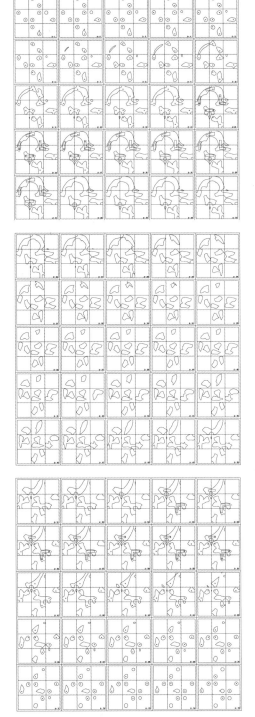

切片图 Drawing for Cutting

第一组 Group 1

制作过程 Fabrication Process

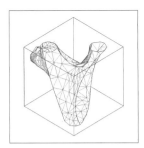

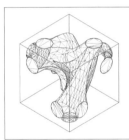
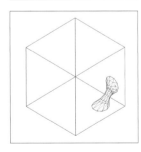
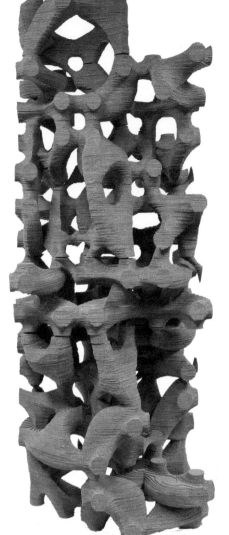

展品名称：冠云
展品材料：软木片
单元种类：4 种
单元总数：59 个
展品尺寸：高 120cm，平面最大长宽各 45cm
设计制作：顾 鹏、胡 博、李沂原、徐臻彦

Title: Guan Yun
Material: cork
Element types: 4
Sum of elements: 59
Dimension: 120cm (height) × 45cm (max. length) × 45cm (max. width)
Designed and fabricated by: GU Peng, HU Bo, LI Yiyuan, XU Zhenyan

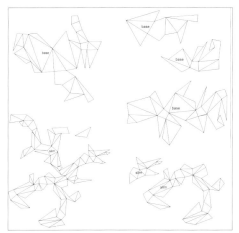
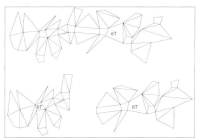
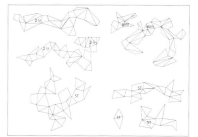

我们尝试了用不同材料诠释设计概念，根据材料本身的属性、连接方式、表现力以及实际制作运输的可能性等方面来做出选择。尝试的材料有：木板、卡纸、玻璃、树脂、竹席纸、穿孔金属网等。最后我们选择了穿孔金属网。在概念表达上，金属网本身的孔隙契合了孔洞的概念主题；其半透明的状态，创造出一种纤薄而敏感的肌质，体现了城市与自然的休戚相关。关于材料本身的属性，金属网便于切割、折叠，空间定位性能好，并且具有一定的强度，不易发生变形。在连接方式上，采用隐匿节点的方法，即通过极其纤细的金属丝把相邻单元的边界缝合一起，形成整体的刚度。另外金属网材料价格低，自重轻，便于实际运输。

To interpret our design concept, we tried different model materials such as timber, cardboard, glass, resin, paper laminated with bamboo matting, perforated metal mesh etc. and evaluated them with the criteria including property, connection way, expression power, the possibility of fabrication and transportation and so on. According to our study on materials, perforated metal mesh was chosen for several reasons. Firstly, the perforation of metal mesh matches the theme – porosity and its translucent appearance as a thin and sensible skin presents a close, interactive relationship between the city and the nature. Furthermore, metal mesh is easy to cut and fold and resistance to distortion. Sewing the edges with tenuous wires makes the connections almost invisible. With low cost and light weight, metal mesh is also easy to transport.

展出模型制作

48

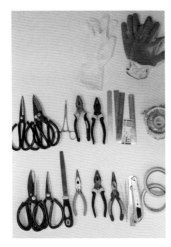
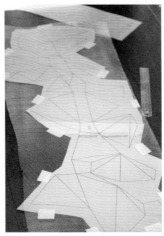

制作过程 Fabrication Process

装配过程 Assembling Process

展出模型制作

50

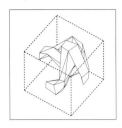
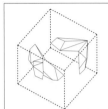
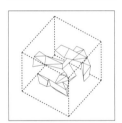
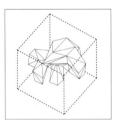
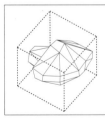
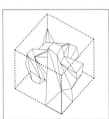

展品名称：绉云
展品材料：金属网
单元种类：6 种
单元总数：54 个
展品尺寸：高 120cm, 平面最大长宽各 75cm
设计制作：曹 婷、杨 晨、杨文杰、周艺南

Title: Zhou Yun
Material: metal mesh
Element types: 6
Sum of elements: 54
Dimension: 120cm (height) × 75cm (max. length) × 75cm (max. width)
Designed and fabricated by: CAO Ting, YANG Chen, YANG Wenjie, ZHOU Yi'nan

第二组 Group 2

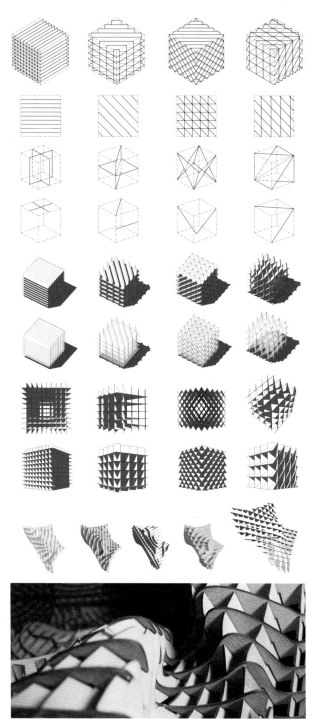

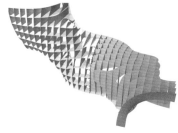

由于采用织构的建造方式，表面肌理的研究显得尤为重要。不同网格的切割方式带来了不同的表面肌理与表面孔洞密度。

Because of the special modeling way – gridding, the study on surface texture becomes crucial. Different cutting ways of the grid influence not only surface texture but also the porosity density of the surface itself.

第三组 Group 3

制作 Fabrication

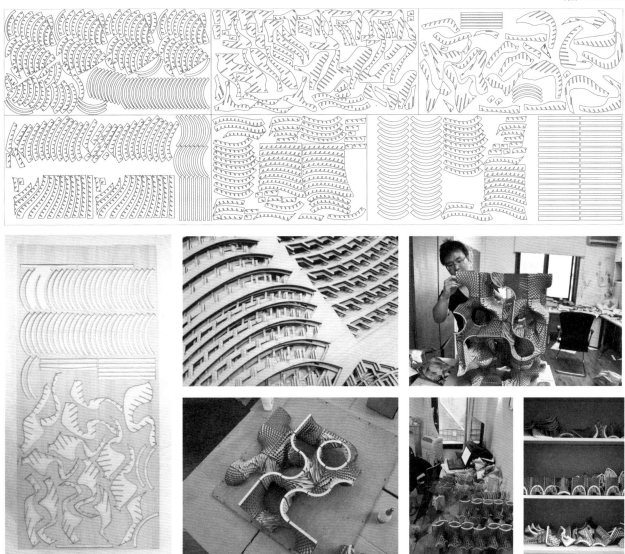

第三组 Group 3

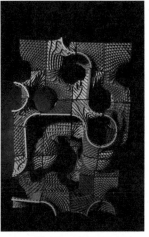
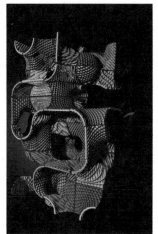

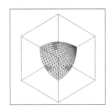
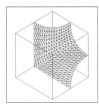
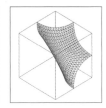
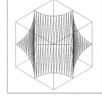

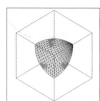

展品名称：瑞云
展品材料：硬木片
单元种类：6 种
单元总数：92 个
展品尺寸：高 72cm,
　　　　　平面最大长宽各 48cm
设计制作：厉鸿凯、李珊珊
　　　　　史 晟、韦栋安

Title: Rui Yun
Material: plywood
Element types: 6
Sum of elements: 92
Dimension:
　72cm (height) × 48cm (max. length) × 48cm (max. width)
Designed and fabricated by:
　LI Hongkai, LI Shanshan, SHI Sheng, WEI Dong'an

冠云峰 Guan-Yun Feng

绉云峰 Zhou-Yun Feng

瑞云峰 Rui-Yun Feng

我们以江南园林中三块著名的石峰命名模型：冠云、绉云、瑞云。它们是中国传统美学中对于孔洞的极致鉴赏。最坚硬实形的物质——石头，在漫长的自然过程中，由柔软无形的物质——水，塑造成形。这不仅是对以透、漏、瘦、皱为美学取向的孔洞形态的欣赏，也是对时间和自然进程的感悟和品鉴。

The three models are entitled Guan-Yun, Zhou-Yun and Rui-Yun, after the names of three most celebrated peculiar stones in Jiangnan Yuanlin (Chinese traditional gardens at the Yangtze Delta). Those odd stones represent the superlative appreciation for porosity in Chinese classical aesthetics. Stone – a hard and solid object is formed gradually over a long period of time in a natural process from water, a soft and formless substance. Those peculiar stones are the incarnation of "tou" (penetrating), "lou" (leaking), "shou" (thinning) and "zhou" (wrinkling). They present not only our admiration for the aesthetics of morphology of porosity, but also our comprehension and appreciation for time and nature course.

 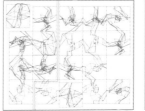 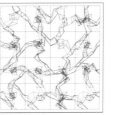

箱体尺寸 830×680×170　　箱体尺寸 830×680×170　　箱体尺寸 640×640×170

装箱与运输
ENCASEMENT & TRANSPORT

实物模型组装

在威尼斯的公寓里，我们只用了两个晚上的时间，将单元体重新组装起来。在贾尔迪尼的明媚阳光下，三件最终完成的模型被安装在奥地利馆内部庭院的指定位置。在这个具有展台和看台双重意义的位置上，它们不仅展示了孔洞作为城市结构的抽象呈现，同时也目睹和见证着第12届威尼斯国际建筑双年展的活动、思想、色彩、情绪、争论和碰撞⋯⋯

ASSEMBLY OF THE EXHIBITION MODELS

Only two nights were needed to assemble the elements in Venice's apartment. Under the sunshine in Giardini, three finished models were anchored (fastened) to their appointed positions in the inner courtyard of the Austrian pavilion. Strategically placed both as a stage and as a stand, the models presented the porosity concept for urban structure at an abstract level. Meanwhile, they brought clarity and bore witness to the vision of the 12th Venice Biennale International Architectural Exhibition with all its activities, concepts, colours, emotions, disputes, and clashes.

奥地利馆庭院中的教学成果展览 Exhibition of Teaching at Courtyard of Austrian Pavillion

城市漫游

城市中的孔洞

孔洞中的城市

模型组装完成后，我们设计了专门的轨迹，让这些孔洞结构在城市中"漫游"，借此观察孔洞与城市空间、其中的人和事件之间的相互关系。在南京，我们选择了东南大学四牌楼校园和具有600年历史的台城城墙；在威尼斯，我们把模型带到了圣马可广场，让它们在这个最负盛名的城市公共空间中表演、寻觅、建立关系、留下印记。

"城市漫游"使我们得以观察孔洞中的城市和城市中的孔洞，在孔洞与城市的交互作用中，反思"孔洞城市"的设计主题。

CITY WANDERING

Porosity within City

City within Porosity

After the models were assembled, a particular wandering track was designed for these porous objects. The path wandered around the cities and allowed observation of the interaction between the structure of porosity and urban space with its people and action. In Nanjing, the models "wandered" both in the campus of Southeast University and on the 600-year-old city wall "Tai Cheng". In Venice, we brought the models to San Marco's Square (Piazza San Marco) and let them perform, seek, establish relationship and leave their footprints in this most respected public space in the world.

"City wandering" enabled us to contemplate "city within porosity", "porosity within city" and the interaction between porosity and city, and turn over to rethink our theme – POROsCITY.

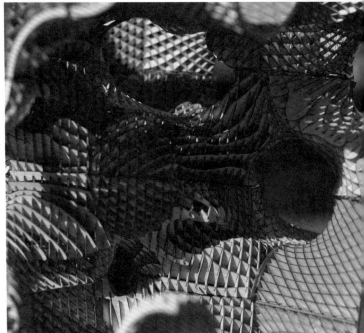

城市漫游 南京 CITY WANDERING NANJING

城市漫游

62

城市漫游　威尼斯　CITY WANDERING　VENICE

城市漫游

城市漫游　威尼斯

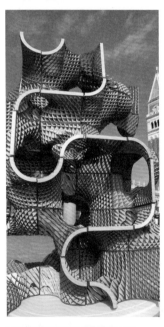

CITY WANDERING   VENICE

在奥地利馆的展出　Exhibition at Austrian Pavillion

展览：第12届威尼斯国际建筑双年展
EXHIBITION: the 12th International Architectural Exhibition / La Biennale di Venezia 2010

第12届威尼斯国际建筑双年展奥地利馆的展览"建造中"于2010年8月27日下午5时在贾尔迪尼双年展公园中开幕。在开幕式上发言的有：奥地利联邦共和国总统海因茨·费舍尔博士；奥地利教育、艺术与文化部部长克劳迪娅·施米德博士；奥地利馆策展人、建筑师艾瑞克·欧文·莫斯。

超过500名各国建筑师、建筑学院教师、艺术家、媒体记者和学生参加了开幕式。

展览延续至2010年11月21日。

At 5pm on August 27th 2010, the exhibition "Under Construction" presented by Austrian Pavilion to the 12th International Architecture Exhibition/La Biennale di Venezia 2010 was opened at the Giardini Biennale. The opening ceremony was addressed by:

Dr. Heinz FISCHER, President of the Federal Republic of Austria; Dr. Claudia SCHMIED, Minister for the Austrian Ministry of Education, Arts and Culture; Eric Owen MOSS, curator of the Austrian Pavilion.

There were altogether 500 people attending the ceremony, including architects, architecture educators, artists, pressmen, and students from around the world.

The exhibition continued until November 21st, 2010.

奥地利馆开幕式 2010年8月27日
Opening Ceremony of Austrian Pavilion 27/08/2010

团队   TEAM

瑞纳·皮尔克
奥地利 rpaX 事务所主持建筑师
东南大学客座教授

Rainer PIRKER
Architect, rpaX Vienna
Guest Professor
Southeast University

张　彤
东南大学建筑学院
教授　建筑系主任

ZHANG Tong
Professor, Director of Architectural Department
School of Architecture
Southeast University

张　慧
东南大学建筑学院
讲师　博士

ZHANG Hui
Lecturer, PhD
School of Architecture
Southeast University

虞　刚
东南大学建筑学院
副教授

YU Gang
Associate Professor
School of Architecture
Southeast University

卡特瑞·塔玛若
奥地利 rpaX 事务所
助理建筑师

Kadri TAMRE
Assistant Architect
rpaX Vienna

顾　鹏
东南大学建筑学院
08 级硕士研究生

GU Peng
Graduate student
School of Architecture
Southeast University

胡　博
东南大学建筑学院
08 级硕士研究生

HU Bo
Graduate student
School of Architecture
Southeast University

李沂原
东南大学建筑学院
09 级硕士研究生

LI Yiyuan
Graduate student
School of Architecture
Southeast University

徐臻彦
东南大学建筑学院
09 级硕士研究生

XU Zhenyan
Graduate student
School of Architecture
Southeast University

曹　婷
东南大学建筑学院
09 级硕士研究生

CAO Ting
Graduate student
School of Architecture
Southeast University

杨　晨
东南大学建筑学院
08 级硕士研究生

YANG Chen
Graduate student
School of Architecture
Southeast University

杨文杰
东南大学建筑学院
09 级硕士研究生

YANG Wenjie
Graduate student
School of Architecture
Southeast University

周艺南
东南大学建筑学院
09 级硕士研究生

ZHOU Yi'nan
Graduate student
School of Architecture
Southeast University

厉鸿凯
东南大学建筑学院
09 级硕士研究生

LI Hongkai
Graduate student
School of Architecture
Southeast University

李珊珊
东南大学建筑学院
09 级硕士研究生

LI Shanshan
Graduate student
School of Architecture
Southeast University

史　晟
东南大学建筑学院
09 级硕士研究生

SHI Sheng
Graduate student
School of Architecture
Southeast University

韦栋安
东南大学建筑学院
08 级硕士研究生

WEI Dong'an
Graduate student
School of Architecture
Southeast University

谨以此书献给瑞纳·皮尔克先生
我至为亲密的师友与合作者
To Rainer PIRKER
My most intimate friend and partner

后 记
Postscript

写这篇后记时，离我们的作品在威尼斯双年展上展出已有七年，离瑞纳·皮尔克先生离开我们也已经六年零四个月了。

"孔洞城市"不只是一次教学，它是一次奇异的研究旅程。自始至终，瑞纳都在驾驶座上。

2010年2月，瑞纳收到奥地利教育、艺术与文化部及策展人艾瑞克·欧文·莫斯的邀请，携其作品参加第12届威尼斯国际建筑双年展奥地利馆的展出。他不加犹豫地选择了东南大学作为合作伙伴，以一次教学活动及其成果，在这个全世界最受瞩目的建筑盛会上展示其对未来城市的理解。

在五个月的教学过程中，我们探究了一种取得城市密度的致密结构，在实质意义上发展了参数化的操作技术和计算方法。这种方法是如此简练，以至于在缺乏CNC工具的情况下，纯粹依靠手工，我们仍然创造出了具有高度艺术性和解释意义的模型，并且可以用最简单和便宜的方式将它们随身携带至展场。

这是一段令所有参与者难忘的经历，尤其是12位研究生，他们在如此年轻的时候，接触并理解着一种前卫的城市理想，经过夜以继日的辛苦工作，将他们的作品展示在国际最高水平的学术目光下。

When I am writing this postscript, it has been seven years since our work was exhibited at Venice Biennale, and six years and four months since Mr. Rainer PIRKER left us.

"POROsCITY" is not only a teaching experience, but also a peculiar journey of researching. Rainer has always sitting in the driver's seat.

In February 2010, Rainer received an invitation from the Ministry of Education, Art and Culture of Austria and from Curator Eric Owen Moss to exhibit his work in Austrian Pavilion at the 12th International Architectural Exhibition in Venice. Without any hesitation, he chose Southeast University as his partner, using the outcomes of a joint teaching studio, to demonstrate our comprehension of future cities at this remarkable architectural feast in front of the whole world.

During a 5-month teaching process, we managed to obtain a concise and compact structure containing urban densities, and developed a parametric operational system and algorithmic method. This method was so succinct that we could generate models conveying highly artistic and interpretative meanings, even without the help of CNC tools, with only basic handwork skills. These models could also be easily and economically transferred into the exhibition site.

This was an unforgettable experience for all of the participants, especially for the 12 graduate students. The students were so young, and they were lucky to engage with and understand this advanced urban ideology. Through many day-and-nights' hard work, they were able to present their work in front of academic sights at the highest international level.

POSTSCRIPT

瑞纳是一位优秀的教师，一位杰出的建筑师，更为重要的是，他是一位真诚、热烈和深刻的理想主义者。我与瑞纳的最初相识，是在2003年"维也纳进步建筑"组织的中奥青年建筑师作品联展，自此我们结下了个人友谊，并开始了学术合作。八年中，我们合作了两次联合教学（第一次是在2004年，名为"都市耕作"，是东南大学最早的国际联合教学之一，成果参加了2004年首届中国国际建筑艺术双年展）。我们还合作完成了天印艺术会馆和南京钟灵街商业综合体的建筑方案设计。

瑞纳有着真诚纯净的内心，他热爱生活，对我们身处的世界有着丰富的见解；他执著于热爱的建筑设计和教育事业，这份执著和敬诚深深感染着我和身边的每一个人；他精湛的专业造诣和开阔的学术视野使我们深受惠泽。

瑞纳是在这份执著中离我们而去的。这本书的付梓，记录了东南大学一次高水平的国际联合教学过程，也是对瑞纳·皮尔克先生的纪念。

邓才德先生与戴天晨博士生为丛书总序及本书的前言、后记的翻译做出贡献，在此深表谢意！

Rainer was an excellent teacher and an outstanding architect, more importantly he was an honest, warm-hearted and deep-thought idealist. We became to know each other in 2003 at the joint exhibition of Chinese-Austrian young architects, organized by Architecture in Progress, Vienna. Since then Rainer and I developed our friendship and academic cooperation. In 8 years we had worked together twice on international joint teaching (The first time was in 2004, titled "Farm the City, Urban Intervention", one of the earliest international joint teaching studios at Southeast University. The result of our work was exhibited at China International Architecture & Art Biennale,2004). We had also worked together on projects such as Heaven Seal Art Gallery and Zhongling Garden of Trees in Nanjing.

Rainer had a pure mind and sincere heart. He loved life and had rich understanding of our world. He was very much obsessed with architecture and architectural education. Myself and people worked around him were always influenced and impressed by his passion and commitment to architecture. We all benefited from his professional competence and broad visions.

Rainer left us with deep obsession with architecture. The publication of this book can be regarded as a record of a unique international joint teaching studio at Southeast University, as well as a commemoration of Mr. Rainer PIRKER.

Thanks to Mr. DENG Caide and doctor candidate Mrs. DAI Tianchen for their contributions to the translation of General Preface, Foreword and Postscript.

张彤
2017.10 于南京

ZHANG Tong
October 2017, Nanjing

图书在版编目（CIP）数据

孔洞城市：第12届威尼斯国际建筑双年展受邀专题设计课程 / 张彤，（奥）瑞纳·皮尔克（Rainer Pirker），张慧著 . -- 南京：东南大学出版社，2017.11
（东南大学建筑学院国际联合教学丛书 / 张彤主编）
ISBN 978-7-5641-7472-9

Ⅰ.①孔… Ⅱ.①张… ②瑞… ③张… Ⅲ.①建筑设计 - 教学研究 - 高等学校 Ⅳ.① TU2

中国版本图书馆 CIP 数据核字（2017）第 269346 号

孔洞城市

责任编辑：	戴　丽　魏晓平
责任印制：	周荣虎
出版发行：	东南大学出版社
社　　址：	南京市四牌楼 2 号（邮编 210096）
网　　址：	http://www.seupress.com
出 版 人：	江建中
印　　刷：	上海雅昌艺术印刷有限公司
开　　本：	889mm×1194mm　1/20
印　　张：	4.5
字　　数：	142 千字
版　　次：	2017 年 11 月第 1 版
印　　次：	2017 年 11 月第 1 次印刷
书　　号：	ISBN 978-7-5641-7472-9
定　　价：	58.00 元
经　　销：	全国各地新华书店
发行热线：	025-83790519　83791830

版权所有，侵权必究
本社图书若有印装质量问题，请直接与营销部联系。电话（传真）：025-83791830